Test Yourself

College Chemistry

Drew H. Wolfe, Ph.D.
Professor of Chemistry
Hillsborough Community College
Tampa, FL

Contributing Editors

Thomas Berke, Ph.D.
Professor of Chemistry
Brookdale Community College
Lincroft, NJ

Paul R. Hemmes, Ph.D.
Vice President of Research and Strategic Planning
Environmental Test Systems, Inc.
Elkhart, IN

C. J. Alexander, Ph.D.
Des Moines Area Community College, Urban Campus
Des Moines, IA

NTC LearningWorks
a division of NTC Publishing Group
Lincolnwood, Illinois

A *Test Yourself Books, Inc.* Project

Published by NTC Publishing Group
© 1996 NTC Publishing Group, 4255 West Touhy Avenue
Lincolnwood (Chicago), Illinois 60646-1975 U.S.A.
All rights reserved. No part of this book may be reproduced, stored
in a retrieval system, or transmitted in any form or by any means,
electronic, mechanical, photocopying, recording, or otherwise, without
prior permission of NTC Publishing Group.
Manufactured in the United States of America.

6 7 8 9 ML 0 9 8 7 6 5 4 3 2 1

Contents

Preface v

How to Use this Book vii

1. Chemistry Fundamentals 1
2. Atoms, Ions, and Molecules 8
3. Moles, Stoichiometry, and Molarity 14
4. Chemical Reactions and Solution Stoichiometry 22
5. Gases and the Kinetic Molecular Theory 30
6. Thermochemistry–The First Law of Thermodynamics ... 39
7. The Electronic Structure of Atoms 49
8. Periodic Properties of Atoms and Elements 57
9. Chemical Bonds–Fundamental Concepts 65
10. The Structure of Molecules and Molecular Orbitals 74
11. Liquids, Solids, and Changes of State 81
12. Solutions and Colloids 91
13. Chemical Kinetics–Rates of Chemical Reactions 100
14. Fundamental Principles of Chemical Equilibria 110
15. Acids, Bases, and Salts 120
16. Acid-Base Equilibrium in Aqueous Solution 128
17. Applications of Aqueous Equilibria 139
18. Chemical Thermodynamics–Entropy, Free Energy, and Equilibria 152
19. Electrochemistry 162
20. Nuclear and Radiation Chemistry 172

Preface

College Chemistry for the *Test Yourself* series has been written as a study aid to help first-year college chemistry students prepare for and achieve higher grades on quizzes, exams, mid-term exams, and final exams. *Test Yourself in College Chemistry* covers all the principal topics in General Chemistry courses including: stoichiometry, physical states, solutions, atomic structure, molecular structure, chemical bonding, chemical kinetics, chemical equilibria, and nuclear chemistry.

Each chapter in *Test Yourself in College Chemistry* has four parts. The first part is a brief discussion of the topic, including important definitions, relationships, and equations. The second part, Test Yourself, consists of both multiple choice and open-format questions and problems that typically appear on General Chemistry exams. You should attempt to answer most of the questions, but pay special attention to those that pertain most directly to your course. After completing the practice test, check your answers against those in the third section, Check Yourself. Besides the correct answers, this section also shows the solutions to numerical problems and explains the answers to nonnumerical problems. When you finish grading the test, fill out the numbers in the Grade Yourself key by circling the numbers of those questions you answered incorrectly and then adding up your misses. This will help you identify those topics you need to review in your textbook and notes. You also might want to retest yourself on these incorrectly answered questions again so that you will not make the same mistakes when you take the actual test.

In conclusion, I would like to thank Fred Grayson who has given me the opportunity to write this and other books.

Drew H. Wolfe, Ph.D.

How to Use this Book

This "Test Yourself" book is part of a unique series designed to help you improve your test scores on almost any type of examination you will face. Too often, you will study for a test—quiz, midterm, or final—and come away with a score that is lower than anticipated. Why? Because there is no way for you to really know how much you understand a topic until you've taken a test. The *purpose* of the test, after all, is to test your complete understanding of the material.

The "Test Yourself" series offers you a way to improve your scores and to actually test your knowledge at the time you use this book. Consider each chapter a diagnostic pretest in a specific topic. Answer the questions, check your answers, and then give yourself a grade. Then, and only then, will you know where your strengths and, more importantly, weaknesses are. Once these areas are identified, you can strategically focus your study on those topics that need additional work.

Each book in this series presents a specific subject in an organized manner, and although each "Test Yourself" chapter may not correspond to exactly the same chapter in your textbook, you should have little difficulty in locating the specific topic you are studying. Written by educators in the field, each book is designed to correspond, as much as possible, to the leading textbooks. This means that you can feel confident in using this book, and that regardless of your textbook, professor, or school, you will be much better prepared for anything you will encounter on your test.

Each chapter has four parts:

Brief Yourself. All chapters contain a brief overview of the topic that is intended to give you a more thorough understanding of the material with which you need to be familiar. Sometimes this information is presented at the beginning of the chapter, and sometimes it flows throughout the chapter, to review your understanding of various *units* within the chapter.

Test Yourself. Each chapter covers a specific topic corresponding to one that you will find in your textbook. Answer the questions, either on a separate page or directly in the book, if there is room.

Check Yourself. Check your answers. Every question is fully answered and explained. These answers will be the key to your increased understanding. If you answered the question incorrectly, read the explanations to *learn* and *understand* the material. You will note that at the end of every answer you will be referred to a specific subtopic within that chapter, so you can focus your studying and prepare more efficiently.

Grade Yourself. At the end of each chapter is a self-diagnostic key. By indicating on this form the numbers of those questions you answered incorrectly, you will have a clear picture of your weak areas.

There are no secrets to test success. Only good preparation can guarantee higher grades. By utilizing this "Test Yourself" book, you will have a better chance of improving your scores and understanding the subject more fully.

Chemistry Fundamentals

 Brief Yourself

Chemistry is the science of matter and the changes it undergoes. Matter is anything that has mass and occupies space. The universe is composed of matter and energy. Energy is the capacity to do work.

A sample of matter is characterized by its chemical and physical properties. A physical property is a characteristic of a particular type of matter that can be measured without changing its composition. A chemical property describes what happens to one type of matter when it changes composition.

Matter can be classified as either being a pure substance or mixture. Elements and compounds are pure substances. Elements are composed of atoms, the smallest unit that retains the chemical properties of the element. Elements combine chemically to produce compounds, which are composed of molecules in covalent compounds and formula units in ionic compounds. Mixtures can be either homogeneous or heterogeneous. A homogeneous mixture, often called a solution, is a physical combination of pure substances that exhibits only one phase (region of matter with distinct boundaries). A heterogeneous mixture is a physical combination of pure substances that exhibits more than one phase.

Matter can exist in three physical states: solids, liquids, and gases. Solids are usually the most dense state of matter, have a fixed shape and volume, and do not exhibit fluid properties. Liquids are usually less dense than solids, have a fixed volume but variable shape, and can flow. Gases are the least dense state of matter, have variable shape and volume, and are the most fluid physical state.

The International System of Units, SI, has seven base units from which we derive all of the remaining units. Prefixes are added to units to scale the unit to the proper size.

Mass is the quantity of matter contained in an object. Volume is the amount of space occupied by matter. The density of matter is the ratio of mass to volume. Temperature is a measure of the degree of hotness of a body, and determines the direction of heat flow.

Test Yourself

1. What is the difference between physical and chemical properties? Give examples of each.

2. What is the difference between intensive and extensive properties? Give examples of each.

3. Classify each the following as intensive or extensive properties. a. volume b. area c. temperature d. boiling point

4. a. Approximately, how many different elements are known to exist? b. What small particles make up elements? c. How do these particles differ from those that make up compounds?

5. Explain how compounds differ from elements?

6. Explain how mixtures differ from pure substances?

7. Aluminum, Al, is a silvery-white metal that melts at 660°C and boils at 2467°C. Al is a good conductor of heat, can be hammered into different shapes, has a low density, readily combines with oxygen to form a thin layer of aluminum oxide, and dissolves in acid, releasing hydrogen gas. Classify each stated property of Al as a physical or chemical property.

8. Classify each of the following as a pure substance or mixture: a. beer b. oxygen c. blood d. carbon e. steel f. distilled water

9. Write the symbols for each of the following elements: a. cobalt b. silver c. fluorine d. sodium e. potassium f. silicon g. mercury

10. Write the names for each of the following elements: a. Cu b. V c. Li d. B e. Br f. Be

11. Write the name and number of each atom in the following: a. H_2SO_4 b. $Mg_3(PO_4)_2$

12. How many atoms of each type are in $Sn_3[Fe(CN)_6]_2$?

13. Explain how a solution of liquid water (bp = 100°C) and liquid ethanol (bp = 78.3°C) could be separated?

14. What physical state best fits each of the following descriptions? a. has the strongest forces among particles b. is the most fluid c. is compressible d. takes the shape of its container to the level it fills

15. At what temperature do each of the following occur? a. liquid changes to a gas b. solid changes to a liquid c. solid changes to a gas

16. List the seven base units of the SI system.

17. Convert 144 km to mm.

18. Normal body temperature is 98.6°F. Convert 98.6°F to degrees Celsius, °C, and to kelvins, K.

19. The melting point of Sb is 630.7°C. Express the melting point of Sb in kelvins.

20. At what temperature is absolute zero in the Fahrenheit scale?

21. How many milligrams, mg, are in one nanogram, ng?

22. One acre is equivalent to an area of 43,560 ft^2. What is the area of 0.50 acre in mm^2?

23. The density of mercury is 13.6 g/cm^3? What is the mass in kg of 255 cm^3 Hg?

24. a. An empty 25.0-cm^3 graduated cylinder has a mass of 53.4 g. The mass of the filled graduated cylinder and an unknown liquid is 88.4 g. What is the density of the unknown liquid? b. Is the liquid in this cylinder water?

25. The density of germanium, Ge, is 5.23 g/cm^3. What is the mass in kg of a cube of Ge that has an edge length, l, of 80.5 mm? ($V_{cube} = l^3$)

26. The density of sulfur dioxide, SO_2, is 2.927 kg/m^3. Calculate the density of sulfur dioxide in g/cm^3.

27. What is the edge length of a cube of gray tin that has a mass of 35.0 g? The density of gray tin is 5.75 g/cm^3.

28. What mass of gold (d_{Au} = 19.3 g/cm^3) occupies the same volume as 10.0 g molybdenum (d_{Mo} = 10.2 g/cm^3)?

29. A light-year is the distance traveled by light in one year. The velocity of light is 3.0×10^8 m/s. Currently the outermost planet in our solar system is Neptune. Neptune is about 2.8 billion miles from the sun. What is the distance in light-years from the sun to Neptune?

30. Describe two factors that introduce uncertainty into measured values.

31. What is the difference between accuracy and precision of measurements?

32. A 50-mL graduated cylinder has 1 mL graduations. What is the maximum number of significant figures of the volumes that can be reported from this graduated cylinder?

33. How many significant figures are indicated by each of the following measurements? a. 11.001 g b. 0.05050 mm^3 c. 0.00004 J d. 11 objects

34. Calculate the average of the following temperatures and express the answer to the correct number of significant figures: 227.6°C, 112.99°C, 21.004°C, and 48°C.

35. Round off each of the following to two significant figures. a. 4.599 mg b. 274.83 mL c. 0.009685 cm

36. Convert 0.78 days to milliseconds, ms, and express the answer to the correct number of significant figures.

37. Perform the following calculation and express the answer to the correct number of significant figures with the correct units.

    8.333 cm × 12 cm =

38. Perform the following calculation and express the answer to the correct number of significant figures with the correct units.

    4.007 g/(12.441 mL – 10.11 mL) =

39. To how many significant figures should the answer to the following be expressed?

    (45.333 g – 44.909 g)/0.5051 mL =

    a. 1
    b. 2
    c. 3
    d. 4
    e. 5

40. How many significant figures are represented in the following measurement?

    0.04000 kg

    a. 3
    b. 4
    c. 5
    d. 6
    e. none of these

Check Yourself

1. Physical properties are characteristics of individual substances and can be measured without reference to any other substance. Chemical properties describe changes in composition. Examples of physical properties are melting point, density, and color. Examples of chemical properties are combustible, explosive, reactive, and inert. **(Matter and its properties)**

2. The magnitude of intensive properties is independent of the amount of substance. The magnitude of extensive depends on the amount present. Intensive properties include melting point and density. Extensive properties include mass and volume. **(Matter and its properties)**

3. a, b–extensive; c, d–intensive **(Matter and its properties)**

4. a. About 110 elements are known. b. Atoms make up elements. c. Molecules are the smallest units of compounds. Molecules are chemical combinations of atoms. **(Classification of matter)**

5. Compounds are composed of elements that are chemically combined; e.g., H_2O and CO_2. **(Classification of matter)**

6. Pure substances have constant compositions, undergo state changes at fixed temperatures, and cannot be separated by physical means. Mixtures have variable compostions, undergo state changes over a range of temperatures, and can be separated by physical means. **(Classification of matter)**

7. Physical properties: melts at 600°C, boils at 2467°C, good conductor of heat, can be hammered into different shapes, has a low density.

    Chemical properties: readily combines with oxygen to form a thin layer of aluminum oxide, dissolves in acid and releases hydrogen gas. **(Classification of matter)**

8. mixtures: a, c, e pure substances: b, d, f **(Classification of matter)**

9. a. Co, b. Ag, c. F, d. Na, e. K, f. Si, g. Hg **(Classification of matter)**

10. a. copper, b. vanadium, c. lithium, d. boron, e. bromine, f. beryllium **(Classification of matter)**

11. a. hydrogen-2, sulfur-1, and oxygen-4, b. magnesium-3, phosphorus-2, and oxygen-8 **(Classification of matter)**

12. tin-3, iron-2, carbon-12, and nitrogen-12 **(Classification of matter)**

13. Distillation, separation by heating, could be used to separated this solution. When heated, the lower boiling ethanol would separate from the higher boiling water. **(Classification of matter)**

14. a. solids, b. gases, c. gases, d. liquids **(States of matter)**

15. a. boiling point, b. freezing point, c. sublimation point **(States of matter)**

16. Mass, kilogram; time, second; length, meter; temperature, kelvin; quantity of substance, mole; electric current, ampere; and luminous intensity, candela **(SI System of Units)**

17. 144 km × (1000 m/1 km) × (1000 mm/1 m) = 1.44×10^8 mm **(SI System of Units)**

18. °C = 5/9(°F − 32) = 5/9(98.6°F − 32) = 37.0°C

    K = °C + 273.2 = 37.0°C + 273.2 = 310.2 K **(SI System of Units)**

Chemistry Fundamentals / 5

19. K = 630.7°C + 273.2 = 903.9 K **(SI System of Units)**

20. °F = 9/5°C + 32 = (9/5 × −273.3°C) + 32 = −459.8°F **(SI System of Units)**

21. 1 ng × (1 g/10^9 ng) × (10^3 mg/1 g) = 10^{-6} mg/ng **(SI System of Units)**

22. 0.50 acre × (43,560 ft^2/1 acre) × (12 in/1 $ft)^2$ × (2.54 cm/1 $in)^2$ × (10 mm/1 $cm)^2$ = 2.02 × 10^9 mm^2 **(SI System of Units)**

23. 255 cm^3 × (1 mL/1 cm^3) × (13.6 g/cm^3) × (1 kg/1000 g) = 3.47 kg **(SI System of Units)**

24. a. 88.4 g cylinder + liquid − 53.4 g cylinder = 35.0 g liquid
    d = mass/volume = 35.0 g/25.0 cm^3 = 1.40 g/cm^3
    b. Water has a density of 1 g/cm^3; thus, this liquid is not water. **(SI System of Units)**

25. V = (80.5 mm × 1 cm/10 $mm)^3$ = 522 cm^3 Ge
    522 cm^3 Ge × 5.23 g/cm^3 × (1 kg/1000 g) = 2.73 kg Ge **(SI System of Units)**

26. 2.927 kg/m^3 × (10^3 g/1 kg) × (1 m/10^2 $cm)^3$ = 2.927 × 10^{-3} g/cm^3 **(SI System of Units)**

27. 35.0 g Sn × (1 cm^3/5.75 g) = 6.09 cm^3
    V = 6.09 cm^3 = l^3
    l = 1.83 cm **(SI System of Units)**

28. 10.0 g Mo × (1 cm^3/10.2 g) = 0.980 cm^3 Mo
    0.980 cm^3 × 19.3 g Au/cm^3 = 18.9 g Au **(SI System of Units)**

29. 2.5 × 10^9 miles × (5280 ft/mi) × (12 in/ft) × (1 m/39.37 in) = 4.5 × 10^{12} m
    365.4 days/yr × (24 hr/day) × (60 min/hr) × (60 s/min) × (3.0 × 10^8 m/s) = 9.5 × 10^{15} m/light year
    4.8 × 10^{12} m × (1 light year/9.5 × 10^{15} m) = 4.8 × 10^{-4} light year **(SI System of Units)**

30. Errors introduce uncertainty in measurements. Systematic errors are errors that can be identified and thus can be corrected. Examples include errors of procedure or calibration. Random errors are those that cannot be identified and thus cannot be corrected. **(Uncertainty in measurements and significant figures)**

31. Accuracy is how close a measurement is to the actual value. Percent deviation is used to measure the degree of accuracy. Precision is how closely grouped repeated measurements are to each other. Significant figures and range of uncertainty are often used to measure precision. **(Uncertainty in measurements and significant figures)**

32. Three significant figures **(Uncertainty in measurements and significant figures)**

33. a. 5, b. 4, c. 1, d. significant figures do not apply to exact measurements **(Uncertainty in measurements and significant figures)**

34. $\dfrac{227.6°C + 112.99°C + 21.004°C + 48°C}{4} = \dfrac{410°C}{4} = 102°C$

    (Uncertainty in measurements and significant figures)

35. a. 4.6 mg, b. 2.7 × 10^2 mL, c. 0.0097 cm **(Uncertainty in measurements and significant figures)**

36. 0.78 day × (24 hr/day) × (60 min/hr) × (60 s/min) × (1000 ms/s) = 6.7 × 10^7 ms **(Uncertainty in measurements and significant figures)**

37. 8.333cm × 12 cm = 99.996 cm^2 which must be rounded to two significant figures; thus the answer is 1.0×10^2 cm^2 **(Uncertainty in measurements and significant figures)**

38. 4.007 g/(12.441 mL − 10.11 mL) = 4.007 g/2.33 mL = 1.72 g/mL **(Uncertainty in measurements and significant figures)**

39. c. The subtraction in the numerator decreases the number of significant figures to three. **(Uncertainty in Measurements and significant figures)**

40. b. Only the zeros after a number are significant. **(Uncertainty in measurements and significant figures)**

Grade Yourself

Circle the numbers of the questions you missed, then fill in the total incorrect for each topic. If you answered more than three questions incorrectly, you need to focus on that topic. (If a topic has less than three questions and you had at least one wrong, we suggest you study that topic also. Read your textbook, a review book, or ask your teacher for help.)

Subject: Chemistry Fundamentals

Topic	Question Numbers	Number Incorrect
Matter and its properties	1, 2, 3	
Classification of matter	4, 5, 6, 7, 8, 9, 10, 11, 12, 13	
States of matter	14, 15	
SI System of Units	16, 17, 18, 19, 20, 21, 22, 23, 24, 25, 26, 27, 28, 29	
Uncertainty in measurement and significant figures	30, 31, 32, 33, 34, 35, 36, 37, 38	

Atoms, Ions, and Molecules

2

Brief Yourself

The first scientific model of the atom was proposed by John Dalton in 1803. The Dalton model could explain the law of conservation of matter, which states that matter cannot be created or destroyed, and the law of constant composition, which states that the composition of a pure substance is constant. Dalton also proposed the law of multiple proportions which states that if two elements can combine to form more than one compound, the masses of one element that combine with a fixed mass of the second element are in a ratio of small whole numbers.

After the discovery of electricity and with experimentation with discharge tubes, scientists found that the atom had a substructure. J.J. Thomson measured the charge-to-mass ratio for the electron and found that the electron had a small mass and high negative charge. R. A. Millikan measured the charge on the electron in the oil-drop experiment. From the results of Thomson's and Millikan's experiments, the mass of the electron was calculated. Research into the nature of radioactivity lead E. Rutherford to the discovery of the nucleus. He showed the existence of the nucleus when he bombarded gold leaf with particles. H. G. J. Moseley used x-rays to measure the charge on the nucleus—the atomic number.

The atomic number, Z, is the number of protons in the nucleus, and the mass number, A, is the total number of protons and neutrons in the nucleus. Isotopes are atoms with the same atomic numbers but have different mass numbers. The atomic mass of an element is the average mass of its naturally occurring isotopes relative to ^{12}C.

The periodic table lists all of the elements. Each element is a member of both a chemical group or family, a vertical column, and a chemical period, a horizontal row. The members of a chemical group share many common properties. Metals are on the left side and nonmetals are on the right side of the periodic table. Metals are usually shiny solids that are good conductors, have high melting and boiling points, and are malleable and ductile. Nonmetals are gases, liquids, or low-melting solids that are usually insulators of heat and electricity.

If an atom loses an electron it becomes a positive ion, a cation. If it gains an electron it becomes a negative ion, an anion. Ions are most stable when they obtain the same number of electrons as noble gases, group 18 (VIIIA). Ions that result when one atom loses or gains electrons are monatomic ions. A charged group of bonded atoms is a polyatomic ion.

Molecules result when two or more nonmetal atoms bond in chemical reactions. The atoms in a molecule are held by chemical bonds. Compounds composed of molecules are known as covalent compounds, and those composed of ionic formula units are ionic compounds. Binary compounds have two different atoms, and ternary compounds have three different atoms. Chemists use the Stock system of nomenclature to assign names to inorganic compounds.

Test Yourself

1. How many neutrons are in the nuclei of $^{204}_{81}\text{Tl}$ atoms?

2. What is the atomic number, mass number, and symbol for the element that has 45 protons and 58 neutrons?

3. Natural samples of copper contain ^{63}Cu and ^{65}Cu. ^{63}Cu has a mass of 62.930 and ^{65}Cu has a mass of 64.928. If the percent abundance of ^{63}Cu is 69.09% and the percent abundance of ^{65}Cu is 30.91%, calculate the atomic mass of copper.

4. Boron is composed of two isotopes, ^{10}B and ^{11}B. The mass of ^{10}B is 10.013 and the mass of ^{11}B is 11.009. If the atomic mass of boron is 10.81, calculate the percent abundances of ^{10}B and ^{11}B.

5. Which of the following are ionic compounds? a. $NaNO_3$, b. H_2S, c. CuI_2, d. CH_3COOH

6. Write the name of SnS_2.

7. Write the name of $Pb_3(PO_4)_2$.

8. Write the formula of titanium(III) oxide.

9. What is the formula of aluminum bisulfite?

10. Predict the formula of the compound that results from O and F.

11. How many elements are in the second period of the periodic table?

12. Three nitrogen oxides are dinitrogen trioxide, dinitrogen tetroxide, and dinitrogen pentoxide. Write the formulas of these compounds.

13. How does a molecule differ from a polyatomic ion? Give an example of each.

14. What scientist proposed the first scientific model of the atom?

15. Pure SiO_2 contains 46.8% Si by mass. A 5.000-g sample of a silicon-oxygen compound is analyzed and found to contain 2.35 g Si. Use the law of constant composition to determine if this sample is pure SiO_2 or is it contaminated with other elements?

16. What contribution did each of the following scientists make to the development of the model of the atom? a. J.J. Thomson b. Robert Millikan c. Ernest Rutherford d. H. G. J. Moseley

17. The charge-to-mass ratio for a p^+ is 1/1837 times that of the charge-to-mass ratio for an e^-, -1.76×10^8 C/g. What does this indicate about the charge and mass of protons compared with those of electrons?

18. What type of radiation (α, β, or γ) can most readily penetrate 0.5 in. of most metals?

19. How many protons, neutrons, and electrons are in $^{26}_{12}\text{Mg}$?

20. Write the symbol of the atom with an atomic number of 53 and a neutron number of 74.

21. What numbers are used to designate each of the following chemical groups? a. chalcogens b. halogens c. noble gases

22. Classify the following elements as metals, nonmetals, or metalloids. a. Si b. Ne c. Ca

23. In what chemical groups on the periodic table do each of the following elements belong? a. Cs b. Sb c. Se

24. Which members of the group 14 (IVA) are classified as nonmetals?

25. List three common properties of the following elements: Ca, Sr, Ba.

26. Which of following elements would you expect to be the best conductor of electricity: Ga, Ge, As, or Se?

27. a. What elements most readily form cations? b. What elements most readily form anions?

28. From their placement on the periodic table, write the expected charge on each of the following ions. a. potassium ions, b. sulfide ions

29. How many electrons are in each of the following ions? a. Hg^{2+} b. P^{3-}

30. Write the names for each of the following polyatomic ions. a. SO_3^{2-} b. NH_4^+ c. CrO_4^{2-}

31. Write the formulas for each of the following. a. nitrite b. bicarbonate c. hypochlorite

32. In what general class of compounds (ionic or covalent) do each of the following belong? a. SCl_2 b. $CaCl_2$ c. $NaNO_3$

33. Explain why it is incorrect to speak of *molecules* of ionic compounds?

34. Write the names for each of the following ternary ionic compounds. a. $Ni(NO_3)_2$ b. $FeSO_3$ c. $Pb(OH)_2$

35. Write the names for each of the following acids. a. HI(*aq*) b. H_2SO_4(*aq*) c. $HBrO_4$(*aq*)

36. Write the names for each of the following. a. $Ba(C_2H_3O_2)_2$ b. $Au(ClO_3)_3$ c. $Sc_2(CrO_4)_3$

37. Write the formulas for each of the following. a. xenon tetrafluoride b. rubidium oxide c. cobalt(III) iodide

38. Write the formulas for each of the following. a. calcium phosphite b. calcium phosphide c. calcium phosphate

39. Write the formulas for each of the following. a. palladium(IV) acetate b. ammonium hypochlorite c. iron(III) cyanide

40. Which of the following scientists discovered that atoms have nuclei?

   a. Rutherford
   b. Thomson
   c. Dalton
   d. Moseley
   e. none of these

Atoms, Ions, and Molecules / 11

Check Yourself

1. 123 n°, $N = A - Z = 204 - 81$ (**Structure of atoms**)

2. $Z = 45$, $A = 103$, ^{103}Rh (**Structure of atoms**)

3. 63.55, Atomic mass = $\dfrac{(62.930 \times 69.09) + (64.928 \times 30.91)}{100} = 63.55$ (**Atomic mass**)

4. 19.98% ^{10}B and 80.02% ^{11}B,
   $x = \%^{10}$B, $y = \%^{11}$B, $x + y = 100$, and $y = 100 - x$
   $x = \%^{10}$B $= 19.7\%$, $y = \%^{11}$B $= 80.3\%$ (**Atomic mass**)

5. a, c. Ionic compounds are composed of a metal and nonmetal or metal and polyatomic ion. (**Compounds**)

6. tin(IV) sulfide   Sulfur in binary compounds has an oxidation number of –2; thus, two S atoms have a total oxidation number of –4. The sum of the oxidation numbers in a compound must equal zero. This means that the charge on Sn is +4. (**Compound names and formulas**)

7. lead(II) phosphate   The charge on phosphate is –3. Two phosphates have a total charge of –6. Hence, the three Pb ions have a charge of +6, and each Pb ion has a charge of +2. (**Compound names and formulas**)

8. Ti_2O_3 All compounds are neutral. To equalize the charges of Ti^{3+} and O^{2-}, the formula must be Ti_2O_3. (**Compound names and formulas**)

9. $Al(HSO_3)_3$ Three HSO_3^- are needed to produce a neutral compound from Al^{3+}. (**Compound names and formulas**)

10. OF_2 An O atom must bond with two F atoms to obtain a stable noble gas configuration. (**Compound names and formulas**)

11. 8 Period 2 consists of the elements from Li to Ne. (**Compound names and formulas**)

12. N_2O_3, N_2O_4, N_2O_5 (**Compound names and formulas**)

13. A molecule is a covalently bonded neutral particle; e.g. CO_2, HCl, H_2O. A polyatomic ion is two or more bonded atoms that carry either a positive or negative charge; e.g. OH^-, CO_3^{2-}, SO_4^{2-}, or NH_4^+.. (**Compounds and ions**)

14. John Dalton (**Structure of atoms**)

15. Because the percent of Si in the compound is 47.0% (2.35 g/5.00 g × 100) and the percent of Si in a pure sample is 46.8%, the sample is fairly pure–within the precision with which the mass measurements were made. (**Structure of atoms**)

16. a. Determined the charge-to-mass ratio of the electron. b. Discovered the charge on an electron. c. Discovered the nucleus. d. Measured the charge on the nucleus. (**Structure of atoms**)

17. Because the charge of a proton is equal but opposite to that of the electron, a charge-to-mass ratio that is 1/1837 that of the electron means that the proton has a mass 1837 times that of the electron. (**Structure of atoms**)

18. Gamma radiation has the greatest penetration power. (**Radiation**)

19. $p^+ = 12$, $n^o = 14$, $e^- = 12$ **(Structure of atoms)**

20. ^{127}I Add the atomic number and neutron number to get the mass number. **(Structure of atoms)**

21. a. VIA or 16, b. VIIA or 17, c. VIIIA or 18 **(Periodic Table)**

22. a. metalloid b. nonmetal c. metal The zigzag staircase separates the metals, on the left, from the nonmetals, on the right. Most of the elements on the staircase are metalloids (semimetals). **(Periodic Table)**

23. a. alkali metals (1) b. group 5 (VA) c. chalcogens (16) **(Periodic Table)**

24. C is the only nonmetal. Si and Ge are metalloids, and the remaining elements are metals. **(Periodic Table)**

25. These three elements are members of the alkaline earth metals (group 20. They are all metals so they are solids, are good conductors of heat and electricity, are malleable and ductile, and have reasonably high melting points. **(Periodic Table)**

26. Ga is the best conductor of electricity because it is a metal. Ge and As are metalloids and Se is a nonmetals. Metals are better conductors than metalloids or nonmetals. **(Periodic Table)**

27. a. Metals most readily lose electrons and form cations. b. Nonmetals most readily gain electrons and form anions. **(Periodic Table)**

28. a. K^+, b. S^{2-} **(Ions)**

29. a. Hg atoms have 80 electrons; thus Hg^{2+} ions have 78 electrons. Cations lose electrons. b. Phosphorus atoms have 15 electrons; thus, P^{3-} ions have 18 electrons. Anions gain electrons. **(Ions)**

30. a. sulfite ion, b. ammonium ion, c. chromate ion **(Compound names and formulas)**

31. a. NO_2^-, b. HCO_3^-, c. OCl^- **(Compound names and formulas)**

32. a. covalent, b. ionic, c. ionic **(Compounds)**

33. Ionic compounds are composed of a network of anions surrounded by cations. This is called a crystal lattice. The simplest ratio of anions to cations is called the formula unit. No molecules are found within this structures. Molecules are associated with covalent compounds. **(Compounds)**

34. a. nickel(II) nitrate b. iron(II) sulfite c. lead(II) hydroxide **(Compound names and formulas)**

35. a. hydroiodic acid b. sulfuric acid c. perbromic acid **(Compound names and formulas)**

36. a. barium acetate b. gold(III) chlorate c. scandium(III) chromate **(Compound names and formulas)**

37. a. XeF_4 b. Rb_2O c. CoI_3 **(Compound names and formulas)**

38. a. $Ca_3(PO_3)_2$ b. Ca_3P_2 c. $Ca_3(PO_4)_2$ **(Compound names and formulas)**

39. a. $Pd(C_2H_3O_2)_4$ b. NH_4OCl c. $Fe(CN)_3$ **(Compound names and formulas)**

40. a. **(Structure of atoms)**

Grade Yourself

Circle the numbers of the questions you missed, then fill in the total incorrect for each topic. If you answered more than three questions incorrectly, you need to focus on that topic. (If a topic has less than three questions and you had at least one wrong, we suggest you study that topic also. Read your textbook, a review book, or ask your teacher for help.)

Subject: Atoms, Ions, and Molecules

Topic	Question Numbers	Number Incorrect
Structure of atoms	1, 2, 14, 15, 16, 17, 19, 20, 40	
Atomic mass	3, 4	
Compounds	5, 32, 33	
Compound names and formulas	6, 7, 8, 9, 10, 11, 12, 30, 31, 34, 35, 36, 37, 38, 39	
Compounds and ions	13	
Radiation	18	
Periodic Table	21, 22, 23, 24, 25, 26, 27	
Ions	28, 29	

Moles, Stoichiometry, and Molarity

3

Brief Yourself

The mole is the counting unit used to keep track of atoms, ions, molecules, and formula units. One mole of any substance has the same number of particles as are in exactly 12 g of ^{12}C. One mole of particles equals 6.022×10^{23} particles, Avogadro's number of particles. One mole of like atoms has a mass equal to their atomic mass in grams, and one mole of molecules has a mass equal to their molecular mass in grams.

The percent composition of a compound is the mass percentage of each element in the compound. The empirical formula expresses the simplest ratio of whole numbers of the atoms in a compound. The molecular formula is the actual ratio of the atoms in a compound.

Chemical equations are used to show what happens in chemical reactions. The substances initially present in a reaction are the reactants and those that form as a result of the reaction are the products. The coefficients in an equation indicate the mole relationships in which the reactants combine and products form. The coefficients in an equation are found by balancing it. Equations are balanced, using the inspection method, by equalizing the number of atoms of each type on either side of the equation.

Stoichiometry is the study of mole, mass, volume, and energy relationships in chemical reactions. Balanced equations allow us to make theoretical predictions concerning the amounts of reactants that combine and the amounts of products that form. The mass of a product predicted from stoichiometric principles is called the theoretical yield. The percent yield of a reaction is the ratio of the actual mass produced to the theoretical yield multiplied times 100. The masses of the products that form depend on the limiting reactant. When the limiting reagent is consumed the reaction stops.

Solutions are homogeneous mixtures of pure substances. A solution is composed of a solute, the dissolved substance, and a solvent, the dissolving substance. Molarity is the ratio of the moles of solute per liter of solution.

Test Yourself

1. What is the mass in grams of 0.115 mol Al?

2. How many gold atoms are in a 10.1-kg sample of pure gold?

3. How many carbon tetrachloride, CCl_4, molecules are in a 0.905-g sample of CCl_4?

4. What is the mass in mg of 4.44×10^{21} formula units of CaF_2?

5. How many F atoms are in a 2.29-g sample of sulfur hexafluoride, SF_6?

6. What mass of KBr has the same number of Br atoms as 5.00 g $MgBr_2$?

7. a. What is the molar mass of titanium?
   b. How many titanium atoms are in a 1.00-mole sample of titanium?

8. What is the molar mass of $CuSO_4 \cdot 5H_2O$?

9. How many moles of lithium atoms are in a 100.0-mg sample of lithium?

10. What is the percent by mass of carbon in $C_6H_{12}O_6$?

11. Calculate the percent composition of potassium permanganate, $KMnO_4$.

12. Calculate the empirical formula of a compound that has the following percent composition: 7.195% P and 92.805% Br.

13. Analysis of an 8.456-g sample of a silicon-chlorine compound shows that it contains 6.519 g chlorine. Calculate the empirical formula of the compound.

14. A sample of a boron-hydrogen compound contains 6.876 g B and 1.069 g H. If the molecular mass of the compound is 74.94, calculate its molecular formula

15. Balance the following equation.

    $B_3N_3H_6 + O_2 \rightarrow N_2O_5 + B_2O_3 + H_2O$

16. Balance the following equation.

    $KMnO_4 + Fe_3O_4 + H_2O \rightarrow Fe_2O_3 + MnO_2 + KOH$

17. Balance the following equation.

    silver sulfide + aluminum + water → silver + aluminum oxide + hydrogen sulfide

18. Consider the following reaction

    $2Cu_2O + Cu_2S \xrightarrow{heat} 6Cu + SO_2$

    What mass of Cu in kg can be obtained when 4.58 kg Cu_2O is heated with excess Cu_2S?

19. Consider the following reaction.

    $Na_2CO_3 + Ca(OH)_2 \rightarrow 2NaOH + CaCO_3$

    What mass of Na_2CO_3 reacts with excess $Ca(OH)_2$ to produce 915 g NaOH

20. Acetylsalicylic acid (aspirin, ASA), $C_9H_8O_4$, is synthesized as follows.

    $C_7H_6O_3 + C_4H_6O_3 \xrightarrow{H_2SO_4} C_9H_8O_4 + C_2H_4O_2$
    Salicylic acid   Acetic anhydride   Aspirin   Acetic acid

    When 10.620 g of salicylic acid (SA), $C_7H_6O_3$, reacts with excess acetic anhydride, $C_4H_6O_3$, in the presence of an acid catalyst, H_2SO_4, 7.762 g ASA is isolated. What is percent yield of the reaction?

21. What mass of HCl can be obtained from 40.3 g of phosgene, $COCl_2$, and 7.49 g of H_2O in the following reaction?

    $COCl_2(g) + H_2O(l) \rightarrow CO_2(g) + 2HCl(g)$

22. Elemental boron, B, results when you heat boron trichloride, BCl_3, to a high temperature in the presence of hydrogen gas, H_2.

$$2BCl_3 + 3H_2 \rightarrow 2B + 6HCl$$

What mass of boron can be obtained from 11.9 g BCl_3 and 0.297 g H_2?

23. What mass of excess reactant remains after the reaction in problem 22 stops?

24. A scientist burns an unknown C, H, and O compound in excess O_2. A 5.000-g sample of this compound produces 9.613 g CO_2 and 3.374 g H_2O. What is the empirical formula of the compound?

25. Methanol is a very toxic alcohol. What is the molarity of a solution that has 7.85 g methanol, CH_4O, dissolved in 153 mL of aqueous solution?

26. How is 150 mL of 0.250 M HNO_3 prepared from concentrated nitric acid, 15.9 M HNO_3?

27. How is a 0.300 M NaBr solution prepared using a 250-mL volumetric flask?

28. What is the molarity of a KI solution that has 13.39 g KI in 450.0 mL of solution?

29. To what total volume must 55.4 mL of 6.00 M HCl be diluted to produce a 1.75 M HCl solution?

30. A chemist dissolves a 1.00-g sample of a mixture of KBr and $MgBr_2$ in H_2O and precipitates the Br^- from solution as AgBr(s) with $AgNO_3$(aq). A precipitate is a solid insoluble substance.

$$Ag^+(aq) + Br^-(aq) \rightarrow AgBr(s)$$

If 1.63 g AgBr is isolated, what is the percent of each component in the mixture?

31. What volume of 12.0 M HCl is needed to prepare 7.35 dm^3 of 0.996 M HCl?

32. Pyruvic acid is a central compound in the production of energy in living cells. When a 2.000-g sample of pyruvic acid is burned in O_2, 2.998 g CO_2 and 0.81844 g H_2O result. What is the empirical formula of pyruvic acid?

33. Concentrated hydrobromic acid, HBr(aq), is sold commercially as a 48% by mass solution and has a density of 1.50 g/cm^3. What is the molarity of this solution?

34. An Al wire has a diameter of 0.0450 cm and a density of 2.67 g/cm^3. What length of this wire in meters contains 1.00 mol Al? (Hint: Assume the Al wire is a cylinder. $V_{cylinder} = \pi r^2 h$)

35. What is the molar mass of $C_{12}H_{22}O_{11}$?

a. 166 g/mol
b. 320 g/mol
c. 342 g/mol
d. 160 g/mol
e. none of these

36. Which of the following contains the largest number of O atoms: 1.00 g O atoms, 1.00 g O_2, or 1.00g O_3?

a. 1.00 g O
b. 1.00 g O_2
c. 1.00 g O_3
d. They have an equal number of O atoms.
e. none of the above

37. What mass of CS_2 has the same number of S atoms as 5.41 g SF_4?

a. 1.91 g
b. 3.81 g
c. 0.0500 g
d. 0.955 g
e. none of these

38. The density of silver is 10.5 g/cm^3. What volume of silver contains 1.00 mol silver atoms?

a. 11.3 L
b. 11.3 mL
c. 0.0952 mL
d. 0.0103 L
e. 10.3 L

39. Morphine, $C_{17}H_{19}NO_3$, is a strong narcotic drug that doctors use as a painkiller. What mass of morphine has the same number of C atoms as 9.0 g CO_2?

    a. 58 g

    b. 3.4 g

    c. 1.5×10^2 g

    d. 0.20 g

    e. none of these

40. What is the percent by mass of zinc in $Zn(ClO_3)_2$?

    a. 33.3%

    b. 23.0%

    c. 30.8%

    d. 25.7%

    e. none of these

Check Yourself

1. 0.115 mol Al × (27.0 g Al/1 mol Al) = 3.11 g **(Mole concept)**

2. 10.1 kg Au × (1000 g/1 kg) × (1 mol Au/197 g Au) × (6.022×10^{23} atoms Au/1 mol) = 3.09×10^{25} atoms Au **(Mole concept)**

3. 0.905 g CCl_4 × (1 mol CCl_4/153.8 g) × (6.02×10^{23} molecules CCl_4/mol) = 3.54×10^{21} molecules CCl_4 **(Mole concept)**

4. 4.44×10^{21} form. units CaF_2 × (1 mol CaF_2/6.02×10^{23} form. units) × (78.1 g CaF_2/mol) × (1000 mg/g) = 575 mg CaF_2 **(Mole concept)**

5. 2.29 g SF_6 × (1 mol SF_6/146.1 g) × (6 mol F/1 mol SF_6) × (6.02×10^{23} atoms F/1 mol) = 5.66×10^{22} atoms F **(Mole concept)**

6. 5.00 g $MgBr_2$ × (1 mol $MgBr_2$/184.1 g) × (2 mol Br/1 mol $MgBr_2$) × (1 mol KBr/1 mol Br) × (119 g KBr/mol) = 6.46 g KBr **(Mole concept)**

7. a. 47.90 g Ti/mol, b. 6.02×10^{23} atoms Ti **(Mole concept)**

8. 249.6 g/mol $CuSO_4 \cdot 5H_2O$ = 159.6 g/mol + 90.0 g/5 mol H_2O **(Mole concept)**

9. 100 mg Li × (1 g/1000 mg) × (1 mol Li/6.94 g) = 0.0144 mol Li **(Mole concept)**

10. %C = (72.0 g C × 180 g $C_6H_{12}O_6$/mol) × 100 = 40.0% C **(Formulas)**

11. %K = (39.10 g K/158.04 g $KMnO_4$/mol) × 100 = 24.74%

    %Mn = (54.94 g Mn/158.04 g $KMnO_4$/mol) × 100 = 34.76%

    %O = (64.00 g O/158.04 g $KMnO_4$/mol) × 100 = 40.50% **(Formulas)**

12. PBr_5

    7.195 g P × (1 mol P/30.97 g) = 0.2323 mol P

    92.805 g Br × (1 mol Br/79.904 g) = 1.615 mol Br

    1.1615 mol Br/0.2323 mol P = 4.999 mol Br/mol P **(Formulas)**

13. Si_3Cl_8

    1.937 g Si × (1 mol Si/28.09 g) = 0.06896 mol Si

    6.519 g Cl × (1 mol Cl/35.45 g) = 0.1839 mol Cl

0.1839 mol Cl/0.06896 mol Si = 2.667 mol Cl/mol Si

Multiply each by 3 (0.667 = 2/3, 2/3 × 3 = 2) to get a ratio of whole numbers.

1.000 mol Si × 3 = 3.00 mol Si

2.667 mol Cl × 3 = 8.001 mol Cl (**Formulas**)

14. B_6H_{10}

    6.876 g B × (1 mol B/10.81 g) = 0.6361 mol B

    1.069 g H × (1 mol H/1.009 g) = 1.061 mol H

    1.061 mol H/0.5361 mol B = 1.668 mol H/mol B

    Multiply each by 3 to get a ratio of whole numbers.

    1.00 mol B × 3 = 3.000 mol B

    1.668 mol H × 3 = 5.004 mol H

    Empirical formula mass = (3 × 10.81) + (5 × 1.008) = 37.47

    Molecular mass/empirical formula mass = 74.94/37.47 = 2

    $B_3H_5 \times 2 = B_6H_{10}$ (**Formulas**)

15. $2B_3N_3H_6 + 15O_2 \rightarrow 3N_2O_5 + 3B_2O_3 + 6H_2O$ (**Equations**)

16. $2KMnO_4 + 6Fe_3O_4 + H_2O \rightarrow 9Fe_2O_3 + 2MnO_2 + 2KOH$ (**Equations**)

17. $3Ag_2S + 2Al + 3H_2O \rightarrow 6Ag + Al_2O_3 + 3H_2S$ (**Equations**)

18. 4.58 kg Cu_2O × (1000 g/kg) × (1 mol Cu_2O/143 g) × (3 mol Cu/1 mol Cu_2O) × (63.5 g Cu/1 mol) × (1 kg/100 g) = 6.10 kg Cu (**Stoichiometry**)

19. 915 g NaOH × (1 mol NaOH/40.0 g) × (1 mol Na_2CO_3/2 mol NaOH) × (106 g Na_2CO_3/1 mol) = 1.21 × 10^3 g Na_2CO_3 (**Stoichiometry**)

20. 10.620 g SA × (1 mol SA/138.12 g) × (1 mol ASA/1 mol SA) × (180.16 g ASA/1 mol) = 13.852 g ASA

    % yield = (7.762 g ASA/13.852 g ASA) × 100 = 56.03% (**Stoichiometry**)

21. 40.3 g $COCl_2$ × (1 mol $COCl_2$/98.9 g $COCl_2$) = 0.407 mol $COCl_2$

    7.49 g H_2O × (1 mol H_2O/18.0 g H_2O) = 0.416 mol H_2O

    Because $COCl_2$ and H_2O react in a one-to-one ratio, $COCl_2$ is the limiting reactant.

    0.407 mol $COCl_2$ × (2 mol HCl/1 mol $COCl_2$) × (36.5 g HCl/1 mol HCl) = 29.7 g HCl (**Stoichiometry**)

22. 11.9 g BCl_3 × (1 mol BCl_3/117.2 g BCl_3) = 0.102 mol BCl_3

    0.297 g H_2 × (1 mol H_2/2.02 g H_2) = 0.147 mol H_2

    Calculate the number of moles of H_2 required to react with 0.102 mol BCl_3.

    0.102 mol BCl_3 × 3 mol H_2/2 mol BCl_3 = 0.153 mol H_2 needed (Thus, H_2 is limiting reactant.)

    0.147 mol H_2 × (2 mol B/3 mol H_2) × (10.8 g B/1 mol B) = 1.06 g B (**Stoichiometry**)

23. 0.147 mol H_2 × (2 mol BCl_3/3 mol H_2) × (117.2 g BCl_3/1 mol BCl_3) = 11.5 g BCl_3

    Mass of excess BCl_3 = 11.9 g – 11.5 g = 0.4 g BCl_3 (**Stoichiometry**)

24. $C_7H_{12}O_4$

    9.613 g CO_2 × (1 mol CO_2/44.01 g) × (1 mol C/1 mol CO_2) × (12.01 g C/1 mol) = 2.623 g C

    3.374 g H_2O × (1 mol H_2O/18.02 g) × (2 mol H/1 mol H_2O) × (1.008 g H/1 mol) = 0.3775 g H

    Total mass = g C + g H + g O

= 5.000 g = 2.623 g C + 0.3775 g H + g O

g O = 5.000 g − (2.623 g + 0.3775 g) = 2.000 g O

Calculate the number of moles of C, H, and O

Mol of C = 2.623 g C × (1 mol C/12.01 g C) = 0.2184 mol C

Mol of H = 0.3775 g H × (1 mol H/1.008 g H) = 0.3745 mol H

Mol of O = 2.000 g O × (1 mol O/16.00 g O) = 0.1250 mol O

Calculate empirical formula

0.2184 mol C/0.1250 mol O = 1.747 mol C/mol O

0.3745 mol H/0.1250 mol O = 2.996 mol H/mol O

0.1250 mol O/0.1250 mol O = 1.000

Multiply each by 4 gives the simplest ratio of whole numbers.

1.747 mol C × 4 = 6.988 mol C = 7 mol C

2.996 mol H × 4 = 11.98 mol H = 12 mol H

1.000 mol O × 4 = 4.000 mol O = 4 mol O **(Combustion analysis)**

25. 7.85 g CH_4O × (1 mol CH_4O/32.0 g) = 0.245 mol CH_4O

 153 mL soln × (1 L soln/1000 mL) = 0.153 L soln

 M = moles CH_4O/L soln = 0.245 mol CH_4O/0.153 L soln = 1.60 M CH_4O **(Molarity)**

26. 150 mL_{dil} × (1 L_{dil}/1000 mL_{dil}) × (0.250 mol HNO_3/L_{dil}) × (1 L_{concd}/15.9 mol HNO_3) ×
 (1000 mL_{concd}/1 L_{concd}) = 2.36 mL 15.9 M HNO_3

 Dilute 2.36 mL of the concentrated acid in enough water to give 150 mL solution. **(Molarity)**

27. 250 mL × (1 L/1000 mL) × (0.300 mol NaBr/L) × (102.9 g NaBr/mol) = 7.72 g

 Dissolve 7.72 g NaBr and dilute to 250 mL. **(Molarity)**

28. 13.39 g KI × (1 mol KI/166.0 g) = 0.08066 mol KI

 450.0 mL × (1 L/1000 m) = 0.4500 L

 M = 0.08066 mol/0.450 L = 0.1793 M **(Molarity)**

29. 55.4 mL × (1L/1000 mL) × (6.00 mol HCl/L) × (1 L/1.75 mol HCl) × (1000 mL/L) = 189 mL

 Dilute the 55.4 mL 6.00 M HCl to 189 mL to produce a 1.75 M HCl solution. **(Molarity)**

30. If x equals the mass of KBr and y equals the mass of $MgBr_2$, then $x + y$ = 1.00 g

 1.63 g AgBr × (1 mol AgBr/188 g) × (1 mol Br/1 mol AgBr) = 8.67 × 10^{-3} mol_{total} Br

 x g KBr × (1 mol KBr/119 g KBr) × (1 mol Br/1 mol KBr) = x/119 mol Br

 y g $MgBr_2$ × (1 mol $MgBr_2$/184 g $MgBr_2$) × (2 mol Br/1 mol $MgBr_2$) = 2y/184 mol Br

 The sum of the moles of Br in each compound equals the total number of moles of Br.

 $$\frac{x}{119} + \frac{2y}{184} = 8.67 \times 10^{-3} \; mol_{total} \; Br$$

 $x + y = 1.00$ g

 $y = 1.00 - x$

 Thus

 $$\frac{x}{119} + \frac{2(1.00 - x)}{184} = 8.67 \times 10^{-3} mol_{total} Br$$

 $x = 0.892$ g, $y = 1.00 - 0.892$ g = 0.108 g

20 / College Chemistry

% KBr = (0.892 g KBr/1.00 g) × 100 = 89.2% KBr

% MgBr$_2$ = (0.108 g MgBr$_2$/1.00 g) × 100 = 10.8 % MgBr$_2$ **(Mixture composition)**

31. 7.35 dm^3 × (1 L/1 dm^3) × (0.996 mol HCl/L) × (1 L/12.0 mol HCl) × (1000 mL/1 L) = 610 mL
    (Molarity)

32. C$_3$H$_4$O$_3$

    2.998 g CO$_2$ × (1 mol CO$_2$/44.01 g) × (1 mol C/1 mol CO$_2$) = 0.06812 mol C × 12.01 g C/mol
    $$= 0.8175 \text{ g C}$$

    0.81844 g H$_2$O × (1 mol H$_2$O/18.016 g) × (2 mol H/1 mol H$_2$O) = 0.090857 mol H × 1.0079 g H/mol
    $$= 0.091575 \text{ g H}$$

    g O = 2.00 g − (0.091575 g H + 0.8175 g C) = 1.0910 g O × (1 mol O/16.00 g) = 0.06819 mol O

    mol C/mol O = 1.00, mol H/mol O = 1.33

    1.00 mol C × 3 = 3 mol C

    1.33 mol H × 3 = 4 mol H

    1.00 mol O × 3 = 3 mol O   **(Combustion analysis)**

33. 48 g HBr/100 g soln × (1 mol HBr/80.9 g) × (1.50 g soln/1 cm^3) × (1000 cm^3/1 L) = 8.9 *M* HBr
    (Molarity)

34. 27.0 g Al/mol × (1 cm^3 Al/2.67 g) = 10.1 cm^3

    V = πr^{2}h = 10.1 cm^3 = 3.14 × (0.0450 cm/2)^2 × h

    h = length = 6.36 × 10^2 cm = 6.36 m   **(Moles)**

35. c. (12 mol × 12 g C/mol) + (22 mol × 1 g H/mol) + (11 mol × 16 g O/mol) = 342 g/mol   **(Mole concept)**

36. d. All have 0.0625 mol O; thus, 1.00 g O in any form has the same number of atoms   **(Mole concept)**

37. a. If they contain the same number of S atoms, than moles of S in each compound are equal, mol S (SF$_4$))
    = mol S CS$_2$.

    5.41 g SF$_4$ × (1 mol SF$_4$/108.1 g) × (1 mol S/1 mol SF$_4$) × (1 mol CS$_2$/2 mol S) × (76.2 g CS$_2$/1 mol)
    = 1.91 g CS$_2$ **(Mole concept)**

38. d. 107.9 g Ag × (1 cm^3/10.5 g Ag) × (1 L/1000 cm^3) = 0.0103 L   **(Mole concept)**

39. b. 9.0 g CO$_2$ × (1 mol CO$_2$/44 g) × (1 mol C/1 mol CO$_2$) × (1 mol C$_{17}$H$_{19}$NO$_3$/17 mol C) × (286 g
    C$_{17}$H$_{19}$NO$_3$/mol) = 3.4 g C$_{17}$H$_{19}$NO$_3$ **(Mole concept)**

40. e. (63.39 g Zn/ × 230.3 g/mol) × 100 = 27.5% Zn   **(Formulas)**

Grade Yourself

Circle the numbers of the questions you missed, then fill in the total incorrect for each topic. If you answered more than three questions incorrectly, you need to focus on that topic. (If a topic has less than three questions and you had at least one wrong, we suggest you study that topic also. Read your textbook, a review book, or ask your teacher for help.)

Subject: Moles, Stoichiometry, and Molarity

Topic	Question Numbers	Number Incorrect
Mole concept	1, 2, 3, 4, 5, 6, 7, 8, 9, 34, 35, 36, 37, 38, 39	
Formulas	10, 11, 12, 13, 14, 40	
Equations	15, 16, 17	
Stoichiometry	18, 19, 20, 21, 22, 23	
Molarity	25, 26, 27, 28, 29, 31, 33	
Mixture composition	30	
Combustion analysis	24, 32	

Chemical Reactions and Solution Stoichiometry

 Brief Yourself

There are four principal classes of inorganic reactions. Combination reactions occur when two or more reactants combine to produce a compound ($A + X \rightarrow AX$). The reverse of a combination reaction is a decomposition reaction in which a compound breaks up and forms elements and/or compounds ($AX \rightarrow A + X$). A single replacement reaction occurs when an element replaces another element in a compound ($AX + B \rightarrow A + BX$). A metathesis reaction is a double replacement reaction ($AX + BY \rightarrow AY + BX$). Examples of metathesis reactions include precipitation, neutralization, and gas-formation reactions.

Electrolytes are substances that either dissociate or ionize to produce ions in aqueous solutions. The opposite of an electrolyte is a nonelectrolyte–it does not produce dissolved ions. Strong electrolytes almost totally break up and form ions, while weak electrolytes only dissociate to a small extent. Ionic equations are written to show the dissolved ions in aqueous reactions. Removing the ions that do not participate in the reaction from an ionic equation produces the net ionic equation.

Precipitation reactions occur when a solid insoluble substance forms in an aqueous reaction. The products of precipitation reactions are predicted using a table of solubilities. Insoluble substances precipitate and soluble ones do not. A titration is a systematic laboratory procedure in which a substance of known concentration is added to an substance of unknown concentration until the equivalence point is reached. The equivalence point is when the number of moles of the unknown reacts completely with the standard solution. In a precipitation titration, a precipitate forms in the titration reaction.

Acids are substances that increase the H^+ concentration in water. Bases are substances that increase the OH^- concentration in water. In neutralization reactions, acids combine with bases and form salts and usually water. In an acid-base titration, a base of known concentration is usually added to an acidic solution of unknown concentration until the moles of OH^- equals the moles of H^+.

Oxidation-reduction reactions are electron-transfer reactions. Substances that lose electrons undergo oxidation. Substances that gain electrons undergo reduction. If a substance undergoes oxidation, it provides the electrons for reduction and is called a reducing agent. If a substance undergoes reduction, it removes electrons and is called an oxidizing agent. A set of signed numbers called oxidation numbers are used to determine if a substance undergoes oxidation or reduction and to help balance some oxidation-reduction equations. Redox equations are balanced using the half-reaction or ion-electron methods.

Test Yourself

1. Write an equation for the dissociation of the strong electrolyte $CaCl_2(s)$ in aqueous solution.

2. Use a solubility table to predict the solubility of the following compounds and state if they precipitate in an aqueous reaction: a. Na_2S b. Hg_2I_2

3. To which of the four classes of inorganic reactions do each of the following belong?

   a. $2CO(g) + O_2(g) \rightarrow 2CO_2(g)$

   b. $Br_2(aq) + 2I^-(aq) \rightarrow 2Br^-(aq) + I_2(aq)$

   c. $PtO_2(s) \rightarrow Pt(s) + O_2(g)$

   d. $H_3PO_4(aq) + 3NaOH(aq) \rightarrow Na_3PO_4(aq) + 3H_2O(l)$

4. Give an example of reaction for each of the following: a. a combination reaction in which oxidation-reduction occurs b. a neutralization reaction using potassium hydroxide c. a single replacement reaction

5. Write the overall and net ionic reactions for the combination of potassium carbonate and copper(II) bromide solutions.

$$K_2CO_3(aq) + CuBr_2(aq) \rightarrow CuCO_3(s) + 2KBr(aq)$$

6. Calculate the molarity of the Cl^- ion in a solution prepared by mixing 100 mL 0.150 *M* NaCl and 100 mL 0.150 *M* $MgCl_2$. Assume the volumes are additive.

7. Using Cl^-, $SO_4^{=}$, and S^{2-} as reagents, outline a procedure for separating $Cu^{2+}(aq)$, $Ca^{2+}(aq)$, and $Pb^{2+}(aq)$.

8. Write the overall and net ionic equation for the following:

   ammonium phosphate(aq) + iron(III) chloride(aq) →

9. What volume of 0.1000 *M* Na_3PO_4 exactly precipitates the Pb^{2+} in 45.35 mL 0.3441 *M* $Pb(NO_3)_2$?

10. What is the difference between a solute that is a weak electrolyte and one that is slightly soluble?

11. What is the difference between the end point and equivalence point in a titration?

12. What is the difference between an Arrhenius acid and a Brønsted-Lowry acid?

13. A 25.00-cm^3 sample of sulfuric acid, H_2SO_4, of unknown concentration is titrated with 0.1219 *M* standard NaOH solution. The equivalence point is reached after adding 18.73 cm^3 of 0.1219 *M* NaOH. What is the molar concentration of the H_2SO_4 solution?

14. A 50.0-mL sample of an unknown phosphoric acid solution, $H_3PO_4(aq)$, is analyzed. When titrated, it takes 39.8 mL 0.0946 *M* NaOH to reach the end point. What is the molarity of the phosphoric acid solution?

15. What is the oxidizing agent in the following reaction?

$$6KOH + 3Br_2 \rightarrow 5KBr + KBrO_3 + 3H_2O$$

16. A 0.5810-g sample of a soluble unknown solid acid is dissolved in water and is titrated to a phenolphthalein end point. If 28.55 mL of 0.1995 *M* NaOH is needed to reach the end point and the solid acid has two moles of H^+ per mole

of acid, what is the molar mass of the unknown acid?

17. A 25.00-mL sample of an acetic acid solution is pipetted and diluted to 100.0 mL. During a titration, 26.99-mL 0.2005 M NaOH is required to reach the equivalence point for a 25.00-mL sample of the diluted acetic acid. If acetic acid releases one H^+ per molecule, what is the concentration of acetic acid in the undiluted solution?

18. What is the oxidation number of sulfur in $Na_2S_2O_3$?

19. What is the oxidation number of manganese in $MnCl_6^{4-}$?

20. Consider the following equation:

$$Cu(s) + 2AgNO_3(aq) \rightarrow Cu(NO_3)_2(aq) + 2Ag(s)$$

What substances undergo oxidation and reduction in this reaction?

21. Identify the oxidizing and reducing agents and what undergoes oxidation and reduction in the following reaction.

$$Cd + NiO_2 + 2H_2O \rightarrow Ni(OH)_2 + Cd(OH)_2$$

22. Identify the oxidizing and reducing agents and what undergoes oxidation and reduction in the following reaction.

$$HCl + HNO_3 \rightarrow NO_2 + \tfrac{1}{2}Cl_2 + H_2O$$

23. Balance the following redox equation.

$$PbS + HNO_3 \rightarrow Pb(NO_3)_2 + NO + S + H_2O$$

24. Balance the following redox equation.

$$As_2O_3 + Cl_2 + H_2O \rightarrow H_3AsO_4 + HCl$$

25. Balance the following redox equation.

$$Al(s) + Ni^{2+}(aq) \rightarrow Ni(s) + Al^{3+}(aq)$$

26. Balance the following redox equation that takes place in an acidic solution.

$$Cr_2O_7{}^{2-}(aq) + Fe^{2+}(aq) \rightarrow Cr^{3+}(aq) + Fe^{3+}(aq) \quad \text{(acid)}$$

27. Balance the following redox equation that occurs in an acidic solution.

$$MnO_4{}^- + Cl^- \rightarrow Mn^{2+} + Cl_2 \quad \text{(acid)}$$

28. Balance the following equation in alkaline solution.

$$S^{2-}(aq) + MnO_4{}^-(aq) \rightarrow S(s) + MnO_2(s) \quad \text{(base)}$$

29. Balance the following redox equation.

$$Al + H_2O \rightarrow Al(OH)_4{}^- + H_2 \quad \text{(base)}$$

30. A 0.8935-g sample of an iron(II) compound is dissolved in water and is titrated with $KMnO_4(aq)$ to the equivalence point. If this titration requires 17.96 mL of 0.02515 M $KMnO_4$ to reach the pink end point, what is the percent by mass of Fe in the sample? (Hint: The products of the reaction are Mn^{2+} and Fe^{3+}.)

31. What are the products of the following reaction?

$$Sr(HCO_3)_2(aq) + HCl(aq) \rightarrow$$

32. Write the balanced oxidation half reaction for the following reaction.

$$H_2O_2(aq) + MnO_4{}^-(aq) \rightarrow O_2(g) + Mn^{2+}(aq) \quad \text{(acid)}$$

33. Ethylene glycol, $C_2H_6O_2$, is oxidized to formaldehyde, H_2CO, by periodic acid, HIO_4. If the HIO_4 changes to $IO_3{}^-$, write a balanced equation for oxidation of ethylene glycol.

34. In which of the following compounds is sulfur in the +5 oxidation state? a. SO_3 b. H_2SO_3 c. SF_2, d. S_8, e. S_2F_{10}

35. Which of the following compounds is a strong electrolyte?
    a. RbCl
    b. NH$_3$
    c. O$_2$
    d. sucrose (table sugar), C$_{12}$H$_{22}$O$_{11}$

36. What is the molar concentration of Cl$^-$ in a 0.15 M CaCl$_2$?
    a. 0.15 M
    b. 0.075 M
    c. 0.30 M
    d. 16.6 M
    e. 33.3 M

37. What volume of 0.500 M K$_2$S is needed to exactly react with 135 mL 0.200 M Al(NO$_3$)$_3$?
    a. 54.0 mL
    b. 108 mL
    c. 27.0 mL
    d. 36.0 mL
    e. 81.0 mL

38. Which of the following is soluble in water?
    a. AgCl
    b. NH$_4$C$_2$H$_3$O$_2$,
    c. CaCO$_3$
    d. FeS,
    e. Co(OH)$_3$

39. What is the net ionic equation for the following?

    Zn(ClO$_3$)$_2$(aq) + Na$_2$S(aq) → ZnS(s) + 2NaClO$_3$(aq)

    a. Na$_2$S → 2Na$^+$ + S^{2-}
    b. ZnS → Zn^{2+} + S^{2-}
    c. Na$^+$ + ClO$_3$– → NaClO$_3$
    d. Zn^{2+} + S^{2-} → ZnS
    e. none of these

40. A 1.098-g sample of a solid acid that contains two mol of H$^+$ per mole of acid is dissolved in water and is titrated with NaOH. If it takes 29.31 cm^3 of 0.2997 M NaOH to neutralize the acid, what is the molar mass of the solid acid?
    a. 250.0 g/mol
    b. 125.0 g/mol
    c. 500.0 g/mol
    d. 0.004392 g/mol
    e. none of these

Check Yourself

1. CaCl$_{2(s)}$ $\xrightarrow{H_2O}$ Ca^{2+}(aq) + 2Cl$^-$(aq) **(Electrolytes)**

2. a. Most sulfides are insoluble compounds, but an exception is the group 1 (IA) sulfides. Therefore, Na$_2$S is soluble and does not precipitate from solution.

   b. Mercury(I), Hg$_2^{2+}$, exists as pairs of metal ions in aqueous solution. Most iodides are soluble compounds, but one of the exceptions is Hg$_2$I$_2$. Thus, Hg$_2$I$_2$ is insoluble and forms a precipitate in aqueous reactions. **(Precipitation)**

3. a. combination (redox), b. single replacement (redox), c. decomposition (redox), d. metathesis **(Inorganic reactions)**

4. a. 2KCl + 3O$_2$ → 2KClO$_3$

   b. KOH + HCl → KCl + H$_2$O

   c. Zn + CuSO$_4$ → ZnSO$_4$ + Cu **(Inorganic reactions)**

5. Because K$_2$CO$_3$, CuBr$_2$, and KBr are strong electrolytes, write them as dissolved ions.

   2K$^+$(aq) + CO$_3^{2-}$(aq) + Cu^{2+}(aq) + 2Br$^-$(aq) → CuCO$_3$(s) + 2K$^+$(aq) + 2Br$^-$(aq)

   After removal of the spectator ions, K$^+$(aq) and Br$^-$(aq), the resulting net ionic equation is as follows:
   Cu^{2+}(aq) + CO$_3^{2-}$(aq) → CuCO$_3$(s) **(Precipitation)**

26 / College Chemistry

6. 100 mL × (1 L/1000 mL) × (0.150 mol NaCl/L) × (1 mol Cl$^-$/mol NaCl) = 0.0150 mol Cl$^-$

    100 mL × (1 L/1000 mL) × (0.150 mol MgCl$_2$/L) × (2 mol Cl$^-$/mol MgCl$_2$) = 0.0300 mol Cl$^-$

    mol Cl$^-$ total = 0.0150 mol + 0.0300 mol = 0.0450 mol

    M(Cl$^-$) = 0.0450 mol/0.200 L = 0.225 M **(Dilution of dissolved ions)**

7. First, separate Pb^{2+} with Cl$^-$, precipitating PbCl$_2$. Then separate Ca^{2+} with SO$_4^{2-}$, by precipitating CaSO$_4$. Finally, remove the Cu^{2+} with S^{2-}, producing solid CuS. **(Precipitation)**

8. 3NH$_4^+$ (aq) + PO$_4^{3-}$(aq) + Fe^{3+}(aq) + 3Cl$^-$(aq) → FePO$_4$(s) + 3NH$_4^+$ (aq) + 3Cl$^-$(aq) **(Overall ionic equation)**

    Fe^{3+}(aq) + PO$_4^{3-}$(aq) → FePO$_4$(s) (Net ionic equation) **(Precipitation)**

9. 3Pb^{2+}(aq) + 2PO$_4^{3-}$ (aq) → Pb$_3$(PO$_4$)$_2$(s)   (net ionic equation)

    45.35 mL Pb^{2+} × (1 L/1000 mL) × (0.3441 mol Pb^{2+}/L) × (2 mol PO$_4^{3-}$/3 mol Pb^{2+}) ×
    (1 mol Na$_3$PO$_4$/1 mol PO$_4^{3-}$) × (1 L/0.1000 mol Na$_3$PO$_4$) = 0.1040 L **(Precipitation)**

10. A weak electrolyte dissociates or ionizes to a small degree but can possibly be very soluble in water. A slightly soluble substance only dissolves to a small degree but can possibly be a strong electrolyte. **(Electrolytes)**

11. At the equivalence point in a titration, the moles of known substance equals the moles of unknown substance. At the end point, the titration is stopped because the indicator has changed color. A good indicator changes color near or at the equivalence point. **(Titrations)**

12. An Arrhenius acid is a substance that increases the H$^+$ concentration in water. A Brønsted-Lowry acid is a substance that donates a proton. A proton and hydrogen ion are exactly the same. The Arrhenius acid needs water as a solvent. The Brønsted-Lowry is independent of the solvent. **(Acid-base)**

13. H$_2$SO$_4$(aq) + 2NaOH(aq) → Na$_2$SO$_4$(aq) + 2H$_2$O(l)

    18.73 cm^3 soln × (1 L soln/1000 cm^3) × (0.1219 mol NaOH/L soln) × (1 mol H$_2$SO$_4$/2 mol NaOH) =
    1.142 × 10^{-3} mol H$_2$SO$_4$

    25.00 mL × (1 L/1000 mL) = 0.025000 L

    $M_{H_2SO_4}$ = 1.142 × 10^{-3} mol H$_2$SO$_4$/0.025000 L soln = 4.566 × 10^{-2} M H$_2$SO$_4$(aq)   **(Acid-base)**

14. H$_3$PO$_4$(aq) + 3NaOH(aq) → Na$_3$PO$_4$(aq) + 3H$_2$O(l)

    39.8 cm^3 soln × (1 L soln/1000 cm^3) × (0.0946 mol NaOH/L soln) × (1 mol H$_3$PO$_4$/3 mol NaOH) =
    1.26 × 10^{-3} mol H$_3$PO$_4$

    50.00 mL × (1 L/1000 mL) = 0.050000 L

    $M_{H_3PO_4}$ = 1.26 × 10^{-3} mol H$_2$SO$_4$/0.050000 L soln = 0.0251 M H$_3$PO$_4$(aq)   **(Acid-base)**

15. Br$_2$ Its oxidation number decreases from 0 to –1, which means it undergoes reduction. **(Redox)**

16. 28.55 mL soln × (1 L soln/1000 mL) × (0.1995 mol NaOH/L soln) × (1 mol OH$^-$/1 mol NaOH) ×
    (1 mol H$^+$/1 mol OH$^-$) × (1 mol acid/2 mol H$^+$) = 2.848 × 10^{-3} mol acid

    MM = g of acid/mol acid = 0.581 g acid/2.848 × 10^{-3} mol acid = 204.0 g acid/mol   **(Acid-base)**

17. 26.99 mL × (1 L/1000 mL) × (0.2005 mol OH$^-$/1 L) × (1 mol diluted AA/1 mol OH$^-$) × (4 mol
    concentrated AA/1 mol diluted AA) = 2.165 × 10^{-2} mol

    M = 2.165 × 10^{-2} mol/0.02500 L = 0.8658 M **(Acid-base)**

18. The charge on the thiosulfate ions is –2 to balance the two positive charge on two sodium ions. The oxida-

tion number for each S is +2. Each O is –2; thus the total for O is –6. Thus the total oxidation number for the two S atoms is +4 because the charge on $S_2O_3^{2-}$ is 2–. **(Redox)**

19. The oxidation number of Mn is +2. The Cl atoms have an oxidation number of –1, giving a total of –6. Hence, Mn is +2 so the oxidation numbers will produce a sum of 4–. **(Redox)**

20. The oxidation state of Cu increases from 0 to +2; thus, it undergoes oxidation. The oxidation state of Ag in $AgNO_3$ decreases from +1 to 0; thus, it undergoes reduction. **(Redox)**

21. Cd increases its oxidation number from 0 to +2 which means that Cd undergoes oxidation and is the reducing agent. Ni in NiO_2 decreases from +4 to +2; thus, NiO_2 undergoes reduction and is the oxidizing agent. **(Redox)**

22. Cl in HCl goes from –1 to 0; thus, HCl undergoes oxidation and is the reducing agent. N in HNO_3 goes from +5 to +4; therefore, HNO_3 undergoes reduction and is the oxidizing agent. **(Redox)**

23. $3PbS + 8HNO_3 \rightarrow 3Pb(NO_3)_2 + 2NO + 3S + 4H_2O$ **(Redox)**

24. $As_2O_3 + 2Cl_2 + 5H_2O \rightarrow 2H_3AsO_4 + 4HCl$ **(Redox)**

25. $2Al(s) + 3Ni^{2+}(aq) \rightarrow 3Ni(s) + 2Al^{3+}(aq)$ **(Redox)**

26. $6Fe^{2+} + 14H^+ + Cr_2O_7^{2-} \rightarrow 2Cr^{3+} + 6Fe^{3+} + 7H_2O$ **(Redox)**

27. $16H^+ + 2MnO_4^- + 10Cl^- \rightarrow 2Mn^{2+} + 5Cl_2 + 8H_2O$ **(Redox)**

28. $3S^{2-} + 2MnO_4^- + 4H_2O \rightarrow 3S + 2MnO_2 + 8OH^-$ **(Redox)**

29. $2OH^- + 6H_2O + 2Al \rightarrow 2Al(OH)_4^- + 3H_2$ **(Redox)**

30. $8H^+(aq) + MnO_4^-(aq) + 5Fe^{2+}(aq) \rightarrow 5Fe^{3+}(aq) + Mn^{2+}(aq) + 4H_2O(aq)$

    17.96 mL MnO_4^- × (1 L MnO_4^-/1000 mL MnO_4^-) × (0.02515 mol MnO_4^-/1 L) × (5 mol Fe^{2+}/1 mol MnO_4^-) × (55.85 g Fe/1 mol Fe) = 0.1261 g Fe

    % Fe = (g Fe/g total) × 100 = (0.1261 g Fe/0.8935 g sample) × 100 = 14.12% Fe   **(Redox)**

31. $SrCl_2(aq)$, $CO_2(g)$, $H_2O(l)$ **(Inorganic reactions)**

32. $H_2O_2(aq) \rightarrow O_2(g) + 2H^+(aq) + 2e^-(aq)$ **(Redox)**

33. $5H_2O + C_2H_6O \rightarrow 2H_2CO_3 + 12H^+ + 12e^-$ (oxidation)

    $2e^- + H^+ + 6HIO_4 \rightarrow IO_3^- + H_2O$ (reduction)

    $C_2H_6O + 6HIO_4 \rightarrow 6IO_3^- + 2H_2CO_3 + 6H^+ + H_2O$ **(Redox)**

34. S_2F_{10} **(Redox)**

35. a. RbCl is a salt that dissociates 100%. **(Electrolytes)**

36. c. 0.15 M $CaCl_2$ × 2 mol Cl^-/1 mol $CaCl_2$ = 0.30 M Cl^- **(Electrolytes)**

37. e. $2Al(NO_3)_3 + 3K_2S \rightarrow 6KNO_3 + Al_2S_3$

    135 mL × (0.200 mol $Al(NO_3)_3$/L) × (3 mol K_2S/2 mol $Al(NO_3)_3$) × (1 L K_2S/0.500 mol) = 81.0 mL
    (Precipitation)

38. b. Both acetates and ammonium compounds are usually soluble; thus, $NH_4C_2H_3O_2$, is the only soluble compound. **(Precipitation)**

39. d. Na^+ and ClO_3^- are spectator ions and are removed from the equation to obtain the net ionic equation. **(Precipitation)**

40. a. 29.31 mL × (1 L/1000 mL) × (0.2997 mol NaOH/L) × (1 mol OH^-/1 mol NaOH) × (1 mol H^+/1 mol OH^-) × (1 mol acid/2 mol H^+) = 0.004392 mol acid

    MM_{acid} = 1.098 g acid/0.004392 mol = 250.0 g acid/mol **(Acid-base)**

Grade Yourself

Circle the numbers of the questions you missed, then fill in the total incorrect for each topic. If you answered more than three questions incorrectly, you need to focus on that topic. (If a topic has less than three questions and you had at least one wrong, we suggest you study that topic also. Read your textbook, a review book, or ask your teacher for help.)

Subject: Chemical Reactions and Solution Stoichiometry

Topic	Question Numbers	Number Incorrect
Electrolytes	1	
Precipitation	2	
Inorganic reactions	3d, 4, 31	
Precipitation	5, 7, 8b, 9, 37, 38, 39	
Dilution of dissolved ions	6	
Overall ionic equation	8a	
Net ionic equation	8b	
Electrolytes	10, 35, 36	
Titrations	11	
Acid-base	12, 13, 14, 16, 17, 40	
Redox	3a, 3b, 3c, 15, 18, 19, 20, 21, 22, 23, 24, 25, 26, 27, 28, 29, 30, 32, 33, 34	

Gases and the Kinetic Molecular Theory

5

Brief Yourself

Gas pressures of isolated samples are measured with a manometer and atmospheric pressure with a barometer. Torr, atm, and kPa are the units of gas pressure most often used. One atm is equivalent to 760 torr and 101.325 kPa.

The ideal gas laws are empirical relationships of the properties of ideal gases. Boyle's law states that the volume of an ideal gas is inversely proportional to its pressure at constant temperature and number of moles. Charles' law states that the volume of a gas is directly proportional to the Kelvin temperature at constant pressure and number of moles. Avogadro's law states that the volume of a gas is directly proportional to the moles of gas at constant temperature and pressure. Avogadro's law implies that equal volumes of ideal gases have the same number of particles. Combining the ideal gas laws gives the ideal gas equation, $PV = nRT$. Dalton's law applies to mixtures of gases, and states that the total pressure of a gaseous mixture equals the sum of the partial pressures of its components. Finally, Gay-Lussac's law of combining volumes states that the volumes of gases in reactions may be expressed as ratios of whole numbers, at constant temperature and pressure.

The kinetic-molecular theory explains the behavior of ideal gases in terms of the motions of the molecules that compose the gas. Five assumptions of the kinetic-molecular theory explain most of the ideal properties of gases. The essence of these assumptions is that the particles in ideal gases move rapidly and randomly throughout their volume without exerting forces on each other. Gas particles essentially occupy none of the total volume; thus, gases are mainly empty space. Collisions among gas particles and between gas particles and the walls of the container are elastic. A wide range of velocities are found for gas particles at a particular temperature, and the average kinetic energy of the particles is proportional to the Kelvin temperature.

At high temperatures and low pressures, gases exhibit properties that approach those of ideal gases. Ideal properties are those predicted by the ideal gas laws. As the temperature decreases and the pressure increases, the properties of gases deviate from ideality. At high pressures, attractive forces cause some molecules to interact and decrease the total pressure exerted by the gas. At even higher pressures, the volume occupied by the gas molecules becomes significant with respect to the total volume, which causes a smaller total volume. The van der Waals equation is used to predict the properties of real gases.

Test Yourself

1. Convert 30.2 inches of Hg, in. Hg, to torr, atmospheres, pascals, and kilopascals.

2. An ideal gas sample enclosed in a 2.35-L bulb at 129 kPa is separated by a closed stopcock from an evacuated 1.91-L bulb. If the temperature is constant, what pressure in atm results when the stopcock is opened and the gas fills both bulbs?

3. A 534-L balloon contains an ideal gas at 25.7°C. At what temperature would this balloon occupy 488 L?

4. A 0.105-g sample of He gas enclosed in a cylinder under a piston occupies 668 cm^3 at 25°C. If 0.0310 g He is removed at constant temperature and pressure, what is the new volume of the cylinder?

5. a. Calculate the value of the ideal gas constant, R, using the molar volume of an ideal gas at STP conditions (1.000 atm and 273.15 K). b. Calculate the value of R in (L·kPa)/(mol·K) using the following relationship: 101.325 kPa/1 atm. c. What is the value of the ideal gas constant in J/(molK)?

6. A sample of N_2 gas occupies 2.64 L at 93.8 kPa and 267 K. What volume does the N_2 occupy at STP conditions?

7. What pressure in torr does 458g N_2 exert in a 43.9-L cylinder at 21.9°C?

8. A 115.7-L cylinder contains $Cl_2(g)$ at 29.1°C and 251.8 kPa. What mass of Cl_2 is in the cylinder?

9. Calculate the density of Rn gas at 298 K and 755 torr.

10. An evacuated 94.751-g glass bulb is filled with a colorless gas. The combined mass of the glass bulb and gas is 95.787 g. To obtain the volume of the flask, it is filled with distilled water. The mass of the flask plus the water is 602.2 g. The pressure of the gas is 833.0 torr and its temperature is 15.9°C. If the density of water is 0.99990 g/cm^3 at 15.9°C, calculate the molar mass of the gas.

11. People exhale a mixture of $N_2(g)$, $O_2(g)$, $CO_2(g)$, and $H_2O(g)$. If the partial pressures of N_2, O_2, and CO_2 in exhaled air are 565, 120, and 28 torr, respectively, and the total pressure is 760 torr, what is the partial pressure of water vapor, P_{H_2O}, in exhaled air?

12. A gaseous mixture of 5.07 g Ar and 9.22 g Kr is in a 6.23-L vessel at 283.8 K. What is the total pressure, P_{total}, of the mixture?

13. A mixture of cyclopropane, C_3H_6, and oxygen, O_2, was once used as a general anesthetic in some medical procedures. Calculate the partial pressures of cyclopropane and oxygen in a mixture that has 0.235 mol C_3H_6 and 0.946 mol O_2 in a 8.11-L container at 298.6 K.

14. The complete combustion of butane, C_4H_{10}, produces $CO_2(g)$ and $H_2O(g)$. If the pressure and temperature remain constant, what volumes of $CO_2(g)$ and $H_2O(g)$ form when 4.58 L C_4H_{10} undergoes combustion?

15. A mixture of $KClO_3$ and MnO_2 is heated, and the resulting $O_2(g)$ is collected over water. If 53.1 mL of O_2 result at 24.0°C and 753.4 torr, what mass of $KClO_3$ did the student heat? The vapor pressure of water at 24.0°C is 22.38 torr.

16. Consider the following gas-phase reaction.

$$2NO(g) + O_2(g) \rightarrow 2NO_2(g)$$

A 1.98-L reaction vessel is filled with a mixture of NO and O_2 to a pressure of 4.33 atm at 310 K. The reaction is allowed to continue until the total pressure decreases to 4.00 atm. What is the final partial pressure of NO_2?

17. A bulb with fixed walls contains $O_2(g)$ at 1.34 atm and 27.8°C. If the bulb is placed into an ice-water bath to 0.0°C. What is pressure of the O_2 in the bulb at 0.0°C?

18. A 68.30-L compressed gas cylinder contains He(g) at 25.9°C and 89.80 atm. How many 2.000-L He balloons at 754.0 torr and 25.9°C can be filled with the He in this cylinder?

19. At 1.00 atm and 273 K, the density of Kr(g) is 3.74 g/L. Calculate the density of Kr at 1.00 atm and −34.1°C.

20. A gaseous mixture contains 2.2 atm He, 1.1 atm H_2, and 4.2 atm N_2. What is the mole fraction of N_2?

21. When ethane, C_2H_6, undergoes complete combustion, it produces $CO_2(g)$ and $H_2O(g)$. What volume of CO_2 results when 229 mL of ethane reacts with excess oxygen at 296 K and 1.04 atm?

22. What names are given to the following ideal gas relationships?

    a. volume and moles at constant temperature and pressure

    b. pressures of gases in mixtures

    c. volume and Kelvin temperature at constant pressure and moles

23. a. What is the value of PV/nRT for all ideal gases? b. Explain why the value of PV/nRT decreases below 1.0 for some real gases at elevated pressures.

24. The van der Waals a constants for He and Cl_2 are 0.034 and 6.49 (atm L^2/mol^2), respectively. The van der Waals b constants for He and Cl_2 are 0.0237 and 0.0562 L/mol, respectively. In terms of their structures, account for the differences in their van der Waals constants.

25. What mass of Ar(g) should be added at constant temperature and pressure to a 5.97-L balloon that contains 10.8 g Ar to increase the volume to 8.23 L?

26. A sample of an unknown gas is allowed to effuse through a small opening in 41.3 min. An equal volume of He gas effuses through the same opening in 9.01 min. Calculate the molar mass of the unknown gas.

27. A 3.20-L sample of $N_2(g)$ is collected over water at 43.0°C and 0.992 atm. After removing the water from $N_2(g)$ sample, it is placed into 1.75-L container at 0.0°C. What pressure does the dry N_2 exert? The vapor pressure of water at 43.0°C is 64.8 torr.

28. A 3.20-g sample of a mixture of NH_4NO_2 and inert KCl is heated. The NH_4NO_2 decomposes to N_2 and H_2O. If 297 mL N_2 is collected over water at 25.0°C and 755 torr, calculate the percent NH_4NO_2 in the mixture. Water's vapor pressure at 25°C is 23.8 torr.

29. A pure gas is composed of 82.66% C and 17.34% H. If the gas has a density of 2.31 g/L at 750 torr and 303 K, calculate the molecular formula of the gas.

30. Compare the root mean square velocities of Kr and Xe molecules at 30.5°C.

31. An unknown gas effuses at a rate 0.632 times the rate of O_2. What is the molar mass of the unknown gas?

32. Write the van der Waals equation and explain the meaning of each variable and constant.

33. a. Use the van der Waals equation to calculate the pressure exerted by 1.00 mol $CO_2(g)$ at 298 K that occupies 65.4 mL. b. Compare the pressure predicted by the van der Waals equation to the pressure obtained from the ideal gas equation for the same conditions and volume. The a and b values for CO_2, are 3.592 $L^2 atm/mol^2$ and 0.0427 L/mol, respectively.

34. Explain in terms of the kinetic-molecular theory of gases why a gas liquefies when cooled sufficiently.

35. If the initial volume of an ideal gas is 39.5 mL at 810 torr, what volume does it occupy at 215.0 kPa?

    a. 19.8 mL

    b. 78.7 mL

    c. 149 mL

    d. 10.5 mL

    e. none of these

36. An ideal gas initially occupies 74.95 mL at 25.3°C. What volume does it occupy at 83.1°C?

    a. 90 mL

    b. 89.46 mL

    c. 246 mL

    d. 62.79 mL

    e. none of these

37. A 0.510-g sample of Ne(g) occupies 495 mL. What volume does the Ne(g) occupy, if 0.250 g Ne is added at constant temperature and pressure?

    a. 332 mL

    b. 495 mL

    c. 243 mL

    d. 739 mL

    e. none of these

38. A sample of Ar gas occupies 90.2 mL at STP conditions. What volume does it occupy at 37°C and 691 torr?

    a. 93.1 mL

    b. 87.4 mL

    c. 113 mL

    d. 13.4 mL

    e. none of these

39. What is the temperature of a 5.05-g sample of Xe(g) that occupies 950 mL at 775 torr?

    a. 306 K

    b. 3.26×10^{-3} K

    c. 2.34 K

    d. 2.34×10^3 K

    e. none of these

40. What volume in mL does a 2.09-g sample of Kr occupy at 925 torr and 333 K?

    a. 559 mL

    b. 3.27×10^{-3} K

    c. 2.34 K

    d. 273 K

    e. none of these

Check Yourself

1. 30.2 in. Hg × (2.54 cm/in) × (10 mm/cm) × (1 torr/1 mmHg) = 767 torr

   767 torr × (1 atm/760 torr) = 1.01 atm

   1.01 atm × (101,325 Pa/atm) = 1.02×10^5 Pa

   1.02×10^5 Pa × (1 kPa/1000 Pa) = 102 kPa (**Pressure**)

2. $P_1V_1 = P_2V_2$

   $P_2 = P_1 \times (V_1/V_2)$ = 129 kPa × (2.35 L/4.26 L) = 71.2 kPa

   71.2 kPa × (1 atm/101.3 kPa) = 0.702 atm (**Combined gas law**)

3. $V_1/T_1 = V_2/T_2$

   $T_2 = T_1 \times (V_2/V_1)$ = 298.9 K × (488 L/534 L) = 273 K = 273 K – 273 = 0°C (**Combined gas law**)

4. n_1 = 0.105 g He × (1 mol He/4.00 g He) = 0.0263 mol He

   n_2 = 0.074 g He × (1 mol He/4.00 g He) = 0.018 mol He

   $V_1/n_1 = V_2/n_2$

   $V_2 = V_1 \times (n_2/n_1)$ = 668 cm³ × (0.018 mol He/0.0263 mol He) = 460 cm³ = 4.6×10^2 cm³
   (**Combined gas law**)

5. a. $R = PV/nT$ = (1.000 atm × 22.414 L)/(1.000 mol × 273.15 K) = 0.08206 (L·atm)/(mol·K)

   b. R = 0.08206 (L·atm)/(mol·K) × 101.325 kPa/atm = 8.314 (L·kPa)/(mol·K)

34 / College Chemistry

c. Pressure times volume, $P \times V$, gives a unit of energy; e.g., $L \times kPa = J$. Thus, the R may be expressed as 8.314 J/(molK). **(Ideal gas law)**

6. $V_2 = 2.64 \text{ L} \times (93.8 \text{ kPa}/101.3 \text{ kPa}) \times (273 \text{ K}/267 \text{ K}) = 2.50 \text{ L}$ **(Combined gas law)**

7. $P = nRT/V = (16.4 \text{ mol} \times 0.0821 \text{ (L·atm)/(mol·K)} \times 295.1 \text{ K})/43.9 \text{ L} = 9.05 \text{ atm}$

    9.05 atm × (760 torr/atm) = 6.88×10^3 torr **(Ideal gas law)**

8. $n = PV/RT = (251.8 \text{ kPa} \times 115.7 \text{ L})/(8.314 \text{ L kPa/mol·K} \times 302.3 \text{ K}) = 11.59 \text{ mol Cl}_2$

    mass = 11.59 mol Cl_2 × (70.90 g Cl_2/mol Cl_2) = 821.7 g Cl_2 **(Ideal gas law)**

9. $P = 755 \text{ torr} \times (1 \text{ atm}/760 \text{ torr}) = 0.993 \text{ atm}$

    d = $(MM \cdot P)/(RT) = (222 \text{ g/mol} \times 0.993 \text{ atm})/(0.0821 \text{ (L·atm)/(mol·K)} \times 298 \text{ K}) = 9.01 \text{ g/L}$
    (Ideal gas law)

10. $m = 95.787 \text{ g} - 94.751 \text{ g} = 1.036 \text{ g}$

    mass of H_2O = 602.2 g – 94.751 g = 507.5 g

    $V_{gas} = V_{H_2O} = 507.5 \text{ g } H_2O \times (1 \text{ mL } H_2O/0.99990 \text{ g } H_2O) \times (1 \text{ L}/1000 \text{ mL}) = 0.5076 \text{ L}$

    $MM = mRT/PV = (1.036 \text{ g} \times 0.08206 \text{ L·atm/K·mol} \times 289.1 \text{ K})/(1.096 \text{ atm} \times 0.5076 \text{ L}) = 44.18 \text{ g/mol}$
    (Ideal gas law)

11. $P_{total} = P_{N_2} + P_{O_2} + P_{CO_2} + P_{H_2O}$

    760 torr = 565 torr + 120 torr + 28 torr + P_{H_2O}

    P_{H_2O} = 760 torr – (565 + 120 + 28) torr = 47 torr **(Dalton's law)**

12. $n_{Ar} = 5.07 \text{ g Ar} \times (1 \text{ mol Ar}/39.95 \text{ g}) = 0.127 \text{ mol Ar}$

    $n_{Kr} = 9.22 \text{ g Kr} \times (1 \text{ mol Kr}/83.80 \text{ g}) = 0.110 \text{ mol Kr}$

    $P_{total} = ((0.0821 \text{ (L·atm)/(mol·K)} \times 283.8 \text{ K})/6.23 \text{ L}) \times (0.127 \text{ mol Ar} + 0.110 \text{ mol Kr}) = 0.886 \text{ atm}$
    (Dalton's law)

13. $P_{total} = RT/V \times n_{total} = RT/V \times (n_{C_3H_6} + n_{O_2}) = ((0.08206 \text{ (L·atm)/(mol·K)} \times 298.6 \text{ K})/8.11 \text{ L}) \times$

    $(0.235 \text{ mol } C_3H_6 + 0.946 \text{ mol } O_2) = 3.568 \text{ atm}$

    $X_{C_3H_6}/n_{total} = 0.235 \text{ mol } C_3H_6/(0.235 \text{ mol } C_3H_6 + 0.946 \text{ mol } O_2) = 0.199$

    $P_{C_3H_6} = X_{C_3H_6} \times P_{total} = 0.199 \times 3.568 \text{ atm} = 0.710 \text{ atm } C_3H_6$

    3.57 atm total – 0.710 atm C_3H_6 = 2.86 atm O_2. **(Dalton's law)**

14. $C_4H_{10}(g) + 13/2 O_2(g) \rightarrow 4CO_2(g) + 5H_2O(g)$

    4.58 L C_4H_{10} × (4 L CO_2/1 L C_4H_{10}) = 18.3 L CO_2

    4.58 L C_4H_{10} × (5 L H_2O/1 L C_4H_{10}) = 22.9 L H_2O **(Stoichiometry)**

15. a. Calculate the moles of O_2 using the ideal gas equation.

    $P_{total} = P_{O_2} + P_{H_2O}$

    753.4 torr = P_{O_2} + 22.38 torr

    P_{O_2} = 731.0 torr × (1 atm/760 torr) = 0.9618 atm

    V = 53.1 mL × (1 L/1000 mL) = 0.0531 L

    T = 24.0°C + 273.2 = 297.2 K

    $PV = n_{O_2}RT$

$n_{O_2} = PV/RT = (0.9618 \text{ atm} \times 0.0531 \text{ L})/(0.08206 \text{ (L·atm)/(mol·K)} \times 297.2 \text{ K}) = 2.09 \times 10^{-3} \text{ mol O}_2$

b. Use the balanced equation and principles of stoichiometry to calculate the mass of $KClO_3$.

$2KClO_3(s) \rightarrow 2KCl(s) + 3O_2(g)$

2.09×10^{-3} mol $O_2 \times$ (2 mol $KClO_3$/3 mol $O_2) \times$ (122.6 g $KClO_3$/1 mol $KClO_3$) = 0.171 g $KClO_3$
(Ideal gas law, Dalton's law)

16. Let x equal the decrease in pressure of O_2.

	$2N_2$ +	O_2 $\rightarrow$	$2NO_2$
Initial	P_{NO}	P_{O_2}	0
Final	$P_{NO} - 2x$	$P_{O_2} - x$	$2x$

$P_{total} = P_{NO} + P_{O_2} + P_{NO_2}$

4.33 atm = $P_{NO} + P_{O_2} + 0$

$P_{total} = P_{NO} + P_{O_2} + P_{NO_2}$

4.00 atm = $(P_{NO} - 2x) + (P_{O_2} - x) + 2x = (P_{NO} + P_{O_2}) - x = 4.33$ atm $- x$

x = 0.33 atm

$P_{NO_2} = 2x = 2 \times 0.33$ atm = 0.66 atm **(Dalton's law)**

17. $P_2 = P_1 \times (T_2/T_1) = 1.34$ atm $\times$ 273.2 K/301.0 K = 1.22 atm **(Combined gas law)**

18. $V_{total} = V_1 \times (P_1/P_2) = 68.30$ L $\times$ (89.90 atm/0.9920 atm) = 6190 L

    6190 L He $\times$ (1 balloon/2 L He) = 3095 balloons **(Combined gas law)**

19. $d_2 = d_1 \times (T_1/T_2)$

    $-34.1°C + 273.2 = 239.1$ K

    $d_2 = 3.74$ g/L $\times$ (273 K/239.1 K) = 4.27 g/L (Combined gas law)

20. P_{total} = 2.2 atm He + 1.1 atm H_2 + 4.2 atm N_2 = 7.5 atm

    X_{N_2} = 4.2 atm N_2/7.5 atm = 0.56 **(Dalton's law)**

21. $C_2H_6 + 3.5O_2 \rightarrow 2CO_2 + 3H_2O$

    229 mL $C_2H_6 \times$ (2 L CO_2/1 L C_2H_6) = 458 mL CO_2 **(Stoichiometry)**

22. a. Avogadro's law, b. Dalton's law of partial pressures, c. Charles' law **(Combined gas laws)**

23. a. $PV = nRT$, $PV/nRT = 1$

    b. The result of the intermolecular forces in real gases is that not all molecules are completely free to move independently. The motion of one influences that of its neighbors and can prevent them from colliding with the walls of the container as hard and as often as in the absence of attractive forces (as in ideal gases). Thus, the pressure is lowered from what would be observed for an ideal gas under the same conditions, which means the product PV becomes smaller than nRT. **(Real gases)**

24. The smaller a value for He indicates that the intermolecular forces among He atoms are smaller than those among Cl_2 molecules. Helium has fewer polarizable electrons than Cl_2; thus, He has weaker intermolecular forces. The smaller b value for He indicates that He atoms take up less volume than Cl_2 molecules. **(Real gases)**

25. 10.8 g Ar $\times$ (1 mol Ar/39.9 g) = 0.271 mol Ar

    $V_1/n_1 = V_2/n_2$

$n_2 = n_1 \times (V_2/V_1) = 0.271$ mol $\times (8.23$ L$/5.97$ L$) = 0.373$ mol

0.373 mol $\times (39.9$ g Ar/mol$) = 14.9$ g

14.9 g $- 10.8$ g $= 4.1$ g Ar **(Combined gas law)**

26. $u_{He}/u_x = (MM_x/MM_{He})^{0.5}$

    The rate of effusion is inversely related to the time; thus, measures of the rate of He and the unknown are expressed as $1/9.01$ min $(0.111$ min$^{-1})$ and $1/41.3$ min $(0.0242$ min$^{-1})$

    0.111 min$^{-1}/0.0242$ min$^{-1} = (MM_x/4.00)^{0.5}$

    $MM_x = 84.2$ g/mol **(Effusion)**

27. 0.992 atm $\times (760$ torr$/1$ atm$) = 754$ torr

    754 torr $- 64.8$ torr $= 689$ torr (dry) $\times (1$ atm$/760$ torr$) = 0.907$ atm

    $P_2 = 0.907$ atm $\times (3.20$ L$/1.75$ L$) \times (273$ K$/316$ K$) = 1.43$ atm **(Combined gas law)**

28. $NH_4NO_2 \rightarrow N_2 + 2H_2O$

    P(dry) $= (755$ torr $- 23.8$ torr$) \times (1$ atm $/760$ torr$) = 0.962$ atm

    $n = PV/RT = (0.962$ atm $\times 0.297$ L$)/(0.0821$ (L·atm)/(mol·K) $\times 298.2$ K$) = 0.0117$ mol N_2

    0.117 mol $N_2 \times (1$ mol $NH_4NO_2/1$ mol $N_2) \times (64.0$ g NH_4NO_2/mol$) = 0.747$ g

    %$NH_4NO_2 = (0.747$ g$/3.20$ g$) \times 100 = 23.3$% **(Stoichiometry)**

29. 82.66 g C $\times (1$ mol C$/12.01$ g$) = 6.883$ mol C$/6.883$ mol C $= 1$

    17.34 g H $\times (1$ mol H$/1.008$ g$) = 17.20$ mol H$/6.883$ mol H $= 2.499 = 2.5$

    Empirical formula $= C_2H_5$, empirical formula mass $= 29$ g/mol

    $PV = nRT$, $PV = g/MM\ RT$, $MM = g/V\ (RT/P)$, $MM = d\ (RT/P)$

    $MM = 2.31$ g/L $\times (0.0821$ (L·atm)/(mol·K) $\times 303$ K$)/0.987$ atm$) = 58.2$ g/mol

    $58.2/29 = 2 \times C_2H_5 = C_4H_{10}$ **(Molecular formula)**

30. $MM_{Kr} = 83.80$ g Kr/mol $\times (1$ kg$/1000$ g$) = 0.08380$ kg Kr/mol

    $MM_{Xe} = 131.3$ g Xe/mol $\times (1$ kg$/1000$ g$) = 0.1313$ kg Xe/mol

    $K = 30.4°C + 273.2 = 303.6$ K

    $$u_a(Kr) = \sqrt{\frac{3 \times 8.314 \frac{\text{kg m}^2}{\text{s}^2} \times 303.6 K}{0.08380 \text{ kg/mol}}} = 301.0 \text{ m/s}$$

    $$u_a(Xe) = \sqrt{\frac{3 \times 8.314 \frac{\text{kg m}^2}{\text{s}^2} \times 303.6 K}{0.1313 \text{ kg/mol}}} = 240.2 \text{ m/s}$$

    The lower molar mass Kr molecules have a higher rms velocity, 301.0 m/s, than the higher molar mass Xe molecules, with a value of 240.2 m/s. **(Effusion)**

31. $\dfrac{\text{rate}_x}{\text{rate}_{O_2}} = \dfrac{0.632}{1}$

    $\dfrac{\text{rate}_x}{\text{rate}_{O_2}} = 0.632 = \sqrt{\dfrac{32.0 \text{g/mol}}{MM_x}}$

$0.632^2 = 0.399 = 32.0$ g/mol/MM_X

$MM_X = 32.0$ g/mol/$0.399 = 80.2$ g/mol **(Effusion)**

32. $\left(P_{obs} + a\left(\dfrac{n}{V}\right)^2\right)(V - nb) = nRT$

    P_{obs} is the observed pressure, V of the container, n is the number of moles of gas, and a and b are the van der Waals constants that correct for deviations from ideality. The pressure correction constant is a, and the volume correction constant is b. **(Real gases)**

33. Because the value for the moles, n, equals 1.00, the van der Waals equation is simplified to

    $\left(P_{obs} + \dfrac{a}{V^2}\right)(V - b) = RT$

    $P_{obs} = (RT/(V - b)) - a/V^2$

    $P_{obs} = (0.08206$ L·atm/mol·K $\times 298$ K/$(0.0654$ L $- 0.0427$ L$)) - 3.592$ L² atm/mol²/$(0.0654$ L$)^2 = 237$ atm

    $P_{ideal} = nRT/V = (1.00$ mol $\times 0.08206$ (L·atm)/(mol·K) $\times 298$ K)/0.0654 L $= 374$ atm

    The ideal gas equation does not consider the attractive forces and the decreased volume. Thus, it gives a higher estimate than the van der Waals equation. **(Real gases)**

34. As the temperature decreases, the average kinetic energy of the particles decrease. At lower energies, the intermolecular forces among the particles are strong enough so the particles can attract and produce the liquid state. **(Kinetic molecular theory)**

35. a. 2.15 kPa $\times$ (1 atm/101.3 kPa) $\times$ (760 torr/1 atm) = 1613 torr

    $V_2 = V_1 \times (P_1/P_2) = 39.5$ mL $\times$ (810 torr/1613 torr) = 19.8 mL **(Combined gas law)**

36. b. $V_2 = V_1 \times (T_2/T_1) = 74.95$ mL $\times$ (356.3 K/298.5 K) = 89.46 mL **(Combined gas law)**

37. d. $V_2 = V_1 \times (n_2/n_1) = 495$ mL $\times$ (0.0376 mol/0.0252 mol) = 739 mL **(Combined gas law)**

38. c. $V_2 = V_1 \times (P_1/P_2) \times (T_2/T_1) = 90.2$ mL $\times$ (760 torr/691 torr) $\times$ (310 K/273 K) = 113 mL **(Combined gas law)**

39. a. $T = PV/nR = (1.02$ atm $\times 0.950$ L)/$(0.0385$ mol $\times 0.0821$ (L·atm)/(mol·K)) = 306 K **(Ideal gas law)**

40. a. $V = nRT/P = (0.0249$ mol $\times 0.0821$ (L·atm)/(mol·K) $\times 333$ K)/1.22 atm = 559 mL **(Ideal gas law)**

Grade Yourself

Circle the numbers of the questions you missed, then fill in the total incorrect for each topic. If you answered more than three questions incorrectly, you need to focus on that topic. (If a topic has less than three questions and you had at least one wrong, we suggest you study that topic also. Read your textbook, a review book, or ask your teacher for help.)

Subject: Gases and the Kinetic Molecular Theory

Topic	Question Numbers	Number Incorrect
Pressure	1	
Combined gas law	2, 3, 4, 6, 17, 18, 19, 22, 25, 27, 35, 36, 37, 38	
Ideal gas law	5, 7, 8, 9, 10, 15, 39, 40	
Dalton's law	11, 12, 13, 15, 16, 20	
Stoichiometry	13, 21, 28	
Real gases	23, 24, 32, 33	
Effusion	26, 30, 31	
Molecular formula	29	
Kinetic molecular theory	34	

Thermochemistry– The First Law of Thermodynamics

Brief Yourself

Thermodynamics is the study of energy and energy transformations. Thermochemistry is the area of thermodynamics concerned mainly with energy changes in chemical reactions.

In thermodynamics, the system is the part of the universe under investigation and everything else is considered the surroundings. The properties of thermodynamic systems are classified as being either state functions or path dependent functions. A state function depends only on the initial and final state and is independent of the path taken to reach the final state. A path dependent function depends on the path taken from the initial to final state. An isothermal system can transfer heat across its walls and thus maintains a constant temperature. An insulated system is called an adiabatic system.

Temperature is a measure of the average kinetic energy of the molecules that make up a system. If two bodies are in thermal equilibrium, then they cannot transfer heat spontaneously from one to the other. If two bodies are in contact and not in thermal equilibrium, then the one with a higher temperature transfers heat to the one at a lower temperature. Heat is a form of kinetic energy that moves from the boundary of a hot object to a colder one. Heat capacity is the heat required to increase the temperature of a substance by 1°C. Molar heat capacity is the heat required to raise one mole of a substance by 1°C. Specific heat capacity is the heat required to raise one gram of substance by 1°C.

A body can possess kinetic energy, potential energy, and internal energy. The internal energy results from the energy possessed by the particles that compose a body. A change in internal energy, E, depends on the heat transfer, q, and work, w, done on or by the system. If heat transfers from the system to the surroundings, then q has a negative value. If heat transfers from the surroundings, then q has a positive value. If work is done by the system on the surroundings, then w has a negative value, and if work is done by the surroundings on the system, then w has a positive value. The sum of q plus w equals the internal energy change for a system.

$$\Delta E = q + w$$

A statement of the first law of thermodynamics is that the energy in the universe is constant. This means that if the system loses energy it enters the surroundings, and if the system gains energy it comes from the surroundings. A mathematical statement of the first law is

$$\Delta E_{sys} + E_{sur} = 0$$

Chemists measure the expansion work done by chemical systems. Work done by a chemical system at constant pressure equals $-P\Delta V$, in which P is the pressure and ΔV is the change in volume. If the volume of a system remains constant, then no work can be done by a chemical system. For constant volume systems, ΔE equals q_v in which q_v is the heat transfer at constant volume. In constant pressure systems, ΔE equals the sum of the heat transfers at constant pressure, q_p, plus the expansion work, $-P\Delta V$. An enthalpy change, H, equals q_p.

Hess's law states that the enthalpy change for a chemical reaction equals the sum of the enthalpy changes of any set of reactions that can be added to give the overall reaction. One of the most common ways to calculate the $\Delta H°$ for a reaction is to use standard enthalpies of formation, $H_f°$. A standard enthalpy of formation is the heat transferred at constant pressure when one mole of a compound forms from its elements at standard conditions (1 atm and 298 K). The $\Delta H°$ for a reaction equals the sum of the standard enthalpies of formation of the products minus the sum of the standard enthalpies of formation of the reactants.

Calorimetry is the procedure for finding the heat transfers in a chemical reaction. A calorimeter is the device used to measure heat transfers. It measures the heat transfer at constant pressure, q_p, or the heat transfer at constant volume, q_v.

Test Yourself

1. Describe each of the following systems: a. isothermal system b. adiabatic system

2. a. What is a state function? b. Give three examples of state functions. c. What is the opposite of a state function?

3. What is the difference between heat and temperature?

4. The temperature increases from 24.6°C to 36.9°C when 24.2 J is added to a 16.9-gram sample of uranium. Calculate the molar and the specific heat capacity of U.

5. How much heat, in J, is needed to raise the temperature of 324 g H$_2$O from 15.3°C to 67.1°C?

6. a. What is the ΔE_{sys} when the system loses 124 J of heat to the surroundings and the system increases in volume by 0.53 L under a pressure of 1.0 atm? b. What ΔE_{sur} accompanies the change in part a?

7. Consider the following reaction:
   $CaCO_3(s) \rightarrow CaO(s) + CO_2(g)$ $\Delta H = +178$ kJ

   a. Is this an exothermic or endothermic reaction?

   b. How many kJ of heat are released/absorbed when 1.00 g calcium carbonate is decomposed?

8. Consider the following reaction:

   $C_3H_8(g) + 5O_2(g) \rightarrow 3CO_2(g) + 4H_2O(l)$
   $\Delta H° = -2220$ kJ

   How much heat is transferred when 1.00 kg C$_3$H$_8$ is burned?

9. Consider the following reaction:

   $\tfrac{1}{2}H_2(g) + \tfrac{1}{2}F_2(g) \rightarrow HF(g)$ $\Delta H_f° = -273$ kJ

   How much heat is transferred when 158 g of HF(g) is produced from the elements?

10. Calculate the $\Delta H°$ for the following reaction:

    $2H_2O_2(l) \rightarrow 2H_2O(l) + O_2(g)$

    Use the following two equations for this calculation:

    $H_2O_2(l) \rightarrow H_2(g) + O_2(g)$ $\Delta H° = +187.8$ kJ

    $H_2(g) + \tfrac{1}{2}O_2(g) \rightarrow H_2O(l)$ $\Delta H° = -285.8$ kJ

11. Consider the following equation:

    $CO(g) + \tfrac{1}{2}O_2(g) \rightarrow CO_2(g)$

The ΔH_f^o values for CO and CO_2 are $\Delta H_f^o =$ −110 kJ and $\Delta H_f^o =$ −393 kJ, respectively. Calculate the enthalpy change for this reaction.

12. Calculate the H^o for the following reaction:

$$P_4O_{10}(s) + 4HNO_3(l) \rightarrow 4H_3PO_4(l) + 2N_2O_5(g)$$

given the following enthalpies of formation:

Compound	ΔH_f^o, kJ/mol
$P_4O_{10}(s)$	−2984
$HNO_3(l)$	−174.1
$H_3PO_4(l)$	−1267
$N_2O_5(g)$	11.3

13. When benzene burns in excess $O_2(g)$, $CO_2(g)$ and $H_2O(g)$ result.

$$C_6H_6(l) + 15/2 O_2(g) \rightarrow 6CO_2(g) + 3H_2O(l)$$

The ΔH_c^o for benzene is −3268 kJ/mol and the ΔH_f^o's for $CO_2(g)$ and $H_2O(l)$ are −394 kJ/mol and −286 kJ/mol, respectively. Calculate the ΔH_f^o for liquid benzene, $C_6H_6(l)$.

14. A calorimeter is calibrated by adding 60.0 g H_2O at 48.9°C to 60.0 g H_2O at 26.2°C in the calorimeter. The final temperature of the H_2O after mixing is 36.2°C. If the specific heat of H_2O is 4.18 J/(g °C), what is the heat capacity, C, for the calorimeter?

15. A calorimeter with a heat capacity of 53.2 J/°C is used to measure the enthalpy of solution of NaOH, $\Delta H^o{}_{soln}(NaOH)$.

$$NaOH \xrightarrow{H_2O} Na^+(aq) + OH^-(aq)$$

Initially, 125 g of H_2O at 24.7°C is transferred to the calorimeter. After adding 5.51 g NaOH to the water, the temperature of the NaOH solution rises to 37.2°C. If the specific heat of the NaOH solution is 3.59 J/(g°C), what is the $\Delta H^o{}_{soln}(NaOH)$ in kJ/mol?

16. a. Explain why it is misleading to say that matter possesses heat. b. How does a body change after heat transfers to it?

17. a. Explain what $w > 0$ means in chemical reactions. b. Explain what $w < 0$ means in chemical reactions.

18. One liter atmosphere, L·atm, is equivalent to 101.3 J. How many kilocalories are equivalent to 500 L·atm?

19. What is the difference between the specific heat and molar heat capacity of a substance?

20. What mass of H_2O can be increased from 30.0 to 50.0°C with the energy released when 93.1 g Cu cools from 84.9°C to 28.3°C? The specific heat of Cu is 0.38 J/(g °C).

21. a. What sign does ΔE have when energy is added to a system? b. What sign does ΔE have when energy is removed from a system?

22. a. What is the ΔE_{sys} for a system that has 325 kJ of work done on it by the surroundings and the system releases 185 kJ of heat? b. What is the ΔE_{sur} for this change?

23. Consider the system in which water vapor condenses to liquid water. Is work done by the system on the surroundings, or is work done by the surroundings on the system, or is no work done?

24. The formation of two moles of NO(g) from nitrogen, N_2, and oxygen, O_2 requires the addition of 180.5 kJ at standard conditions. a. Write the equation for the formation of NO. b. What is the ΔH^o for the reaction? c. Is this reaction an endothermic reaction or an exothermic reaction? Explain.

25. The reaction of Al and Fe_2O_3 is called the thermite reaction. This reaction releases enough energy to weld metals.

$$2Al + Fe_2O_3 \rightarrow 2Fe + Al_2O_3 \quad \Delta H^o = -851.4 \text{ kJ}$$

Calculate the heat produced when 1.00 g Al combines with excess Fe_2O_3.

26. Acetylene, C_2H_2, and O_2 combine and produce CO_2, H_2O, and a large quantity of energy.

    $$C_2H_2(g) + 5/2 O_2(g) \rightarrow 2CO_2(g) + H_2O(g)$$
    $$\Delta H = -1.30 \times 10^3 \text{ kJ}$$

    What mass of acetylene is needed to combine with excess oxygen to increase the temperature of 1.0 kg H_2O from 15.5°C to 83.5°C, if only 87.1% of heat released is absorbed by the water?

27. a. Write a sentence that states Hess's law.
    b. Write a mathematical statement of Hess's law.

28. Calculate the $\Delta H°$ for

    $$OF_2(g) + H_2O(g) \rightarrow O_2(g) + 2HF(g)$$

    given the following data:

    $\frac{1}{2}O_2(g) + F_2(g) \rightarrow OF_2(g)$ $\Delta H° = +23.0$ kJ

    $H_2(g) + \frac{1}{2}O_2(g) \rightarrow H_2O(g)$ $\Delta H° = -241.8$ kJ

    $HF(g) \rightarrow \frac{1}{2}H_2(g) + \frac{1}{2}F_2(g)$ $\Delta H° = +268.6$ kJ

29. A mixture of methane, $CH_4(g)$, and oxygen, $O_2(g)$, is added to a cylinder under a piston. Initially, the volume of the cylinder is 575 cm³. The mixture is then ignited and the products $CO_2(g)$ and $H_2O(g)$ form. If the reaction produces 1.95 kJ of energy and all of the energy is converted to work, what is the final volume of the cylinder under a constant pressure of 1.05 atm? Assume that only the products remain after the reaction.

30. A bomb calorimeter is used to measure the $\Delta H°_c$ of hydrogen gas, $H_2(g)$. The bomb is filled with 0.133 g H_2 and excess O_2. The calorimeter contains 722 g H_2O ($C_{H_2O} = 4.184$ J/(g°C)) initially at 21.44°C. After the reaction the temperature of the H_2O is 26.40°C. If the heat capacity of the calorimeter is 812 J/°C, what is the $\Delta H°_c$ for H_2?

31. In the early 19th-century Dulong and Petit observed that the product of the atomic mass and the specific heat in cal/(g°C) equals roughly 6.4. If the specific heat of Pb is 0.031 cal/(g°C), calculate the atomic mass of Pb using the Delong and Petit relationship.

32. The molar enthalpy of vaporization, ΔH_{vap}, is the enthalpy change when one mole of a liquid at a fixed temperature changes to vapor. The ΔH_{vap} of H_2O at 100°C is 40.6 kJ/mol. Calculate the amount of heat required to change 10.0 g $H_2O(l)$ at 25°C to $H_2O(g)$ at 100°C.

33. Given the following standard enthalpies of formation,

Compound	$\Delta H°_f$, kJ/mol
$Na_2CO_3(s)$	−1131
$HCl(aq)$	−167
NaCl	−407
$H_2O(l)$	−286
$C_2(g)$	−394

    calculate the $\Delta H°$ for

    $$Na_2CO_3(s) + 2HCl(aq) \rightarrow 2NaCl(aq) + H_2O(l) + CO_2(g)$$

34. A simple styrofoam calorimeter contains 125 g H_2O at 23.2°C. A 87.5-g sample of a metal at 100°C is placed in the water. The temperature of the water and metal rise to a maximum of 26.1°C. What is the specific heat of the metal?

35. When 364 J is added to a 55.7-g sample of titanium, the temperature increases from 30.5°C to 42.9°C. Calculate the molar heat capacity of titanium.

    a. 25.2 J/(mol °C)

    b. 47.9 J/(mol °C)

    c. 0.527 J/(mol °C)

    d. 62.7 J/(mol °C)

    e. none of these

36. The specific heats of gold, silver, copper, and water are 0.13, 0.23, 0.38, and 4.184 J/(g °C), respectively. Considering 1.0-g samples of these substances, which would require the most heat to increase their temperature from 273 K to 298 K?

    a. Au

    b. Ag

    c. Cu

    d. H_2O

    e. all would require the same amount of heat

37. Which type of energy is most important in chemical thermodynamics?

    a. kinetic energy

    b. potential energy

    c. internal energy

    d. work

    e. none of these

38. How much work, in kJ, is expended when a gas expands from 5.0 L to 15.0 L under a constant pressure of 2.00 atm?

    a. −20.0 kJ

    b. +20.0 kJ

    c. −2.03 kJ

    d. +2.03 × 10^3 kJ

    e. none of these

39. A system is compressed 3.16 L by the surroundings under a constant pressure of 10.0 atm, and the system releases 1.9 kJ of heat to the surroundings. What is ΔE_{sys}?

    a. 3.20 kJ

    b. −5.10 kJ

    c. 5.10 kJ

    d. 1.3 kJ

    e. none of these

40. Glucose, $C_6H_{12}O_6(s)$, is one of the principal fuels that cells use to produce energy. The enthalpy of combustion of glucose is −2816 kJ. The equation for this reaction is as follows.

    $$C_6H_{12}O_6(s) + 6O_2(g) \rightarrow 6CO_2(g) + 6H_2O(l)$$

    A nutrition table shows that the recommended energy intake for a male between the ages of 23 and 50 years is 11.3 MJ/day. What mass of glucose produces 11.3 MJ of heat?

    a. 4.01 g

    b. 722 g

    c. 2.03 × 10^3 g

    d. 120 g

    e. none of these

Check Yourself

1. a. An isothermal system maintains a constant temperature. b. An adiabatic system does not allow heat to escape or enter the system. **(Thermodynamic terms)**

2. a. A state function only depends on the initial and final states. b. Pressure, temperature, and volume are examples of state functions. c. The opposite of a state function is a path function. **(Thermodynamic terms)**

3. Heat is the form of kinetic energy that is spontaneously transferred from a warmer to a colder body. Temperature is a measure of the average kinetic energy of the particles. **(Thermodynamic terms)**

4. c = 24.2 J/(16.9 g U × (36.9°C − 24.6°C)) = 0.116 J/(g °C)

   C = 0.116 J/(g °C) × (238 g U/mol U) = 27.6 J/(mol °C) **(Heat capacity)**

5. $q = m (T_2 - T_1) c = 324$ g $H_2O \times (67.1°C - 15.3°C) \times 4.18$ J/(g°C) $= 7.02 \times 10^4$ J (70.2 kJ)
   (Heat capacity)

6. a. $\Delta E_{sys} = q + w$

   $w = -P \Delta V = -1.0$ atm $\times$ (+0.53 L) $= -0.53$ L·atm

   $w = -0.53$ L·atm $\times$ 101.3 J/L·atm $= -54$ J

   $\Delta E_{sys} = -124$ J + (–54 J) = –178 J

   b. $\Delta E_{sys} + \Delta E_{sur} = 0$

   $\Delta E_{sur} = -\Delta E_{sys} = -(-178$ J) = +178 J **(Energy transfers)**

7. a. Endothermic, because the enthalpy change is positive.

   b. 1.00 g $CaCO_3 \times$ (1 mol $CaCO_3$/100.1 g) $\times$ (+178 kJ/mol $CaCO_3$) = +1.78 kJ **(Stoichiometry)**

8. 1.00 kg $C_3H_8 \times$ (1000 g C_3H_8/1 kg C_3H_8) $\times$ (1 mol C_3H_8/44.1 g C_3H_8) $\times$ (–2220 kJ/molC_3H_8)
   $= -5.03 \times 10^4$ kJ **(Stoichiometry)**

9. 158 g HF $\times$ (1 mol HF/20.0 g HF) $\times$ (–273 kJ/mol HF) = -2.16×10^3 kJ **(Stoichiometry)**

10. $2H_2O_2(l) \rightarrow 2H_2(g) + 2O_2(g)$ $\quad \Delta H° = +375.6$ kJ (2 × +187.8 kJ)

    $+ 2H_2(g) + O_2(g) \rightarrow 2H_2O(l)$ $\quad \Delta H° = -571.6$ kJ (2 × –285.8 kJ)

    $2H_2O_2(l) \rightarrow 2H_2O(l) + O_2(g)$ $\quad \Delta H° = -196.0$ kJ **(Hess's law)**

11. $\Delta H° = \Sigma \Delta H_f°$(products) $- \Sigma \Delta H_f°$(reactants) $= \Delta H_f°(CO_2) - [\Delta H_f°(CO) + \Delta H_f°(O_2)] = -393$ kJ –
    [(–110 kJ) + 0 kJ] = –283 kJ **(Hess's law)**

12. $\Delta H° = \Sigma \Delta H_f°$(products) $- \Sigma \Delta H_f°$ (reactants)

    = [(4 mol $\times \Delta H_f°(H_3PO_4)) +$ (2 mol $\times \Delta H_f°(N_2O_5))] - [\Delta H_f°(P_4O_{10}) +$ (4 mol $\times \Delta H_f°(HNO_3))]$

    = [(4 mol $\times$ (–1267 kJ/mol) + (2 mol $\times$ (–11.3 kJ/mol)] – [–2984 kJ + (4 mol $\times$ –174.1 kJ/mol)]
    = –1365 kJ **(Hess's law)**

13. $\Delta H_c°(C_6H_6) = $ (6 mol $\Delta H_f°(CO_2) + 3$ mol $\Delta H_f°(H_2O)) - (\Delta H_f°(C_6H_6))$

    –3268 kJ = (6 mol (–394 kJ/mol) + 3 mol (–286 kJ/mol) – $\Delta H_f°$ (C_6H_6)

    $\Delta H_f°$ (C_6H_6) = –46 kJ/mol **(Hess's law)**

14. $q_{lost} = -q_{gained}$

    $q_{hot\,H_2O} = -(q_{cold\,H_2O} + q_{calorimeter})$

    $q_{hot\,H_2O} = m_{hotH_2O} \Delta T_{hotH_2O} c_{H_2O} = 60.0$ g $\times$ (36.2°C – 48.9°C) $\times$ 4.18 J/(g°C) = -3.19×10^3 J

    $q_{cold\,H_2O} = m_{coldH_2O} \Delta T_{coldH_2O} c_{H_2O} = 60.0$ g $\times$ (36.2°C – 26.2°C) $\times$ 4.18 J/(g°C) = 2.51×10^3 J

    $q_{calorimeter} = (36.2°C - 26.2°C) \times c_{calorimeter} = 10.0$ °C $\times c_{calorimeter}$

    $q_{hot\,H_2O} + q_{cold\,H_2O} + q_{calorimeter} = 0$

    -3.19×10^3 J + 2.51×10^3 J + (10.0°C $\times c_{calorimeter}$) = 0

    10.0°C $\times c_{calorimeter} = 6.80 \times 10^2$ J

    $C_{calorimeter} = 6.80 \times 10^2$ J/10.0 °C = 68 J/°C **(Heat capacity and calorimetry)**

15. $q = -(q_{NaOH(aq)} + q_{calorimeter})$

    $q_{NaOH(aq)} = m_{soln} \times \Delta T \times c_{NaOH(aq)}$

    $= (125 \text{ g H}_2\text{O} + 5.51 \text{ g NaOH}) \times (37.2°C - 24.7°C) \times 3.59 \text{ J/(g°C)} = 5.86 \times 10^3 \text{ J}$

    $= 5860 \text{ J}$

    $q_{calorimeter} = \Delta T_{calorimeter} \, C_{calorimeter} = (37.2°C - 24.7°C) \, 23.5 \text{ J/°C} = 294 \text{ J}$

    $q = -(q_{NaOH(aq)} + q_{calorimeter}) = -(5860 \text{ J} + 294 \text{ J}) = -6154 \text{ J}$

    $-6154 \text{ J} \times 1 \text{ kJ}/1000 \text{ J} = -6.154 \text{ kJ}$

    $5.51 \text{ g NaOH} \times (1 \text{ mol NaOH}/40.0 \text{ g NaOH}) = 0.138 \text{ mol}$

    $\Delta H°_{soln}(\text{NaOH}) = -6.154 \text{ kJ}/0.138 \text{ mol} = -44.6 \text{ kJ/mol NaOH}$ (**Calorimetry**)

16. a. Matter cannot possess heat. Heat is usually only detected when bodies at different temperature contact each other. Heat flows spontaneously from the warmer to the colder body.

    b. The average kinetic energy of the particles in a body that receives heat increases. (**Thermodynamic terms**)

17. a. When work is positive ($w > 0$), the volume of the products, V_{prods}, is smaller than that of the reactants, V_{react}. This means that the surroundings has done work on the system (reaction). The system compresses, $V_{prods} < V_{react}$.

    b. When work is negative, the volume of the products is larger than that of the reactants. This means that the system (reaction) has done work on the surroundings. The system expands. (**Thermodynamic terms**)

18. $500 \text{ L atm} \times (101.3 \text{ J/L atm}) \times (1 \text{ cal}/4.184 \text{ J}) \times (1 \text{ kcal}/1000 \text{ cal}) = 12.1 \text{ kcal}$ (**Work and heat**)

19. Specific heat is the amount of heat needed to raise the temperature of one gram of substance by one degree Celsius (which is equivalent to 1 K). Molar heat capacity is the amount of heat needed to raise the temperature of one mole of substance by one degree Celsius. (**Specific heat**)

20. $q_{Cu} = 93.1 \text{ g} \times (28.3°C - 84.9°C) \times 0.38 \text{ J/(g °C)} = -2.00 \times 10^3 \text{ J}$

    $q_{H_2O} = -q_{Cu} = 2.00 \times 10^3 \text{ J} = g_{H_2O} \times (50.0°C - 30.0°C) \times 4.184 \text{ J/(g °C)}$

    $g_{H_2O} = 23.9 \text{ g}$ (**Energy transfers**)

21. a. ΔE is positive. b. ΔE is negative. (**Thermodynamic terms**)

22. a. $\Delta E = q + w = -185 \text{ kJ} + 325 \text{ kJ} = +140 \text{ kJ}$

    b. $\Delta E_{sys} + \Delta E_{sur} = 0$, $\Delta E_{sur} = -\Delta E_{sys}$, $\Delta E_{sur} = -(+140 \text{ kJ}) = -140 \text{ kJ}$ (**Work and heat**)

23. When a gas, water vapor, condenses to a liquid, water, the system has compressed ($V_f < V_i$). This means that work was done on the system by the surroundings. This is always the case when V less than zero. (**Thermodynamic terms**)

24. a. $N_2(g) + O_2(g) \rightarrow 2NO(g)$,

    b. $\Delta H = +180.5 \text{ kJ}$,

    c. This is an endothermic reaction because the enthalpy increases–heat is absorbed in this reaction. (**Stoichiometry**)

25. $1.00 \text{ g Al} \times (1 \text{ mol Al}/27.0 \text{ g}) \times (851.4 \text{ kJ}/2 \text{ mol Al}) = 15.8 \text{ kJ}$ (**Stoichiometry**)

46 / College Chemistry

26. $q_{H_2O} = 1.0 \times 10^3$ g $\times$ (83.5°C − 15.5°C) × 4.184 J/(g °C) × (100 J total/87.1 J) = 3.3 × 10² kJ

3.3×10^2 kJ × (1 mol C_2H_2/1.3 × 10² kJ) × (26 g C_2H_2/mol) = 6.5 g C_2H_2 **(Stoichiometry)**

27. a. Hess's law states that the enthalpy change for a chemical reaction equals the sum of the enthalpy changes of any set of reactions that can be added to give the overall reaction.

   b. $\Delta H° = \Sigma H°_f$(products) − $\Sigma H°_f$(reactants) **(Hess's law)**

28. $OF_2(g) \rightarrow \frac{1}{2}O_2(g) + F_2(g)$ $\Delta H° = -23.0$ kJ

   $H_2O(g) \rightarrow H_2(g) + \frac{1}{2}O_2(g)$ $\Delta H° = 241.8$ kJ

   $+H_2(g) + F_2(g) \rightarrow 2HF(g)$ $\Delta H° = -268.6$ kJ × 2

   $OF_2(g) + H_2O(g) \rightarrow O_2(g) + 2HF(g)$ $\Delta H° = -318.4$ kJ **(Hess's law)**

29. $w = -P\Delta V = -1.95 \times 10^3$ J × (1 L·atm/101.3 J) = −1.05 atm × (V_f − 0.575 L)

   $\Delta V = 18.9$ L   **(Work and heat)**

30. $\Delta T = 26.40°C − 21.44°C = 4.96°C$

   $q_{bomb} = 4.96°C \times 812$ J/°C = 4.03×10^3 J

   $q_{H_2O} = 722$ g × 4.96 °C × 4.184 J/(g °C) = 1.50×10^4 J

   $\Delta H°_c/g = -(q_{bomb} + q_{H_2O})/g\ H_2 = -(4.03 \times 10^3$ J + 1.50×10^4 J$)/0.133$ g = -1.43×10^5 J/g

   $\Delta H°_c = -1.43 \times 10^5$ J/g × (2.02 g H_2/mol) × (1 kJ/1000 J) = −289 kJ/mol   **(Calorimetry)**

31. Atomic mass = 6.4/0.031 = 2.1×10^2 **(Specific heat)**

32. q = 10.0 g × 75°C × 4.184 J/(g °C) = 3.1 kJ

   q = 10.0 g H_2O × (1 mol H_2O/18.0 g) × 40.6 kJ/mol = 22.6 kJ

   q_{total} = 3.1 kJ + 22.6 kJ = 25.7 kJ   **(Energy transfers)**

33. $\Delta H° = \Sigma\Delta H°_f$(products) − $\Sigma\Delta H°_f$ (reactants)

   = [(2 mol × $\Delta H°_f$((NaCl)) + $\Delta H°_f(H_2O)$) + $\Delta H°_f(CO_2)$] − [$\Delta H°_f(Na_2CO_3)$ + (2 mol × $\Delta H°_f$(HCl)]

   = (2 mol × (−407 kJ/mol) + (−286 kJ) + (−394 kJ)] − (−1131 kJ + 2 mol × (−167 kJ/mol)] =
                                                                       −29 kJ   **(Hess's law)**

34. $c_{metal} = -(125$ g $H_2O \times (26.1°C − 23.2°C) \times 4.184$ J/(g °C))/(87.5 g × (23.2°C − 100°C) =
                                                                           0.23 J/(g °C)   **(Specific heat)**

35. a. C_{Ti} = 364 J/(55.7 g × 1 mol Ti/47.9 g × 12.4°C) = 25.2 J/(mol °C) **(Specific heat)**

36. d. The substance with the highest specific heat, H_2O, will require the most heat to raise the temperature of a fixed mass of substance by the same change in temperature. **(Specific heat)**

37. c. The internal energy is the energy associated with the particles that compose matter. **(Thermodynamic terms)**

38. c. $w = -P\Delta V =$ = −2.00 atm × (15.0 L − 5.0 L) × (101.3 J/Latm) × (1 kJ/1000 J) = −2.03 kJ **(Work and heat)**

39. d. $w = -P\Delta V = = -10.0$ atm $\times -3.16$ L $\times$ (101.3 J/Latm) $\times$ (1 kJ/1000 J) $= -3.20$ kJ

   $\Delta E_{sys} = q + w = -1.9$ kJ $+ 3.20$ kJ $= 1.3$ kJ **(Work and heat)**

40. b. 11.3 MJ $\times$ (1000 kJ/MJ) $\times$ (1 mol $C_6H_{12}O_6$/2816 kJ) $\times$ (180 g $C_6H_{12}O_6$/mol $C_6H_{12}O_6$) $= 722$ g $C_6H_{12}O_6$ **(Stoichiometry)**

Grade Yourself

Circle the numbers of the questions you missed, then fill in the total incorrect for each topic. If you answered more than three questions incorrectly, you need to focus on that topic. (If a topic has less than three questions and you had at least one wrong, we suggest you study that topic also. Read your textbook, a review book, or ask your teacher for help.)

Subject: Thermochemistry: The First Law of Thermodynamics

Topic	Question Numbers	Number Incorrect
Thermodynamic terms	1, 2, 3, 16, 17, 21, 23, 37	
Heat capacity	4, 5, 14	
Energy transfers	6, 20, 32	
Stoichiometry	7, 8, 9, 24, 25, 26, 40	
Hess's law	10, 11, 12, 13, 27, 28, 33	
Calorimetry	14, 15, 30	
Work and heat	18, 22, 29, 38, 39	
Specific heat	19, 31, 34, 35, 36	

The Electronic Structure of Atoms

 ## Brief Yourself

The fundamental principle of quantum theory is that the energy released or absorbed by small particles is in discrete units called quanta. A quantum is a discrete packet of energy. One of the first scientists to apply quantum theory was Einstein when he explained the photoelectric effect. The photoelectric effect occurs when light hits the surface of a metal and ejects electrons.

Diffracting the light emitted by a gas in discharge tubes produces a line spectrum, with specific wavelengths separated by dark regions, or a continuous spectrum, with a wide range of wavelengths. Analysis of the line spectrum of hydrogen by Balmer revealed that a mathematical equation could be written to predict the wavelengths of the lines in the visible region as a function of integer values.

Bohr used some of the principles of quantum theory to develop a model of the atom that could explain the spectrum of hydrogen. His model of the H atom placed the electron in energy states which he called stationary states. When in these states the H atom could not emit or absorb energy. Bohr stated that the H atom only emits or absorbs energy when it moves from one stationary state to another. The frequency of the quanta released by atoms is calculated using Planck's equation, $\Delta E = h\nu$. Bohr also believed that the electron in a H atom traveled in concentric circular orbits around the nucleus. The Bohr model was replaced because it could only explain single-electron systems and new scientific evidence showed that some parts of the basic postulates of the Bohr theory were incorrect.

de Broglie proposed that electrons exhibit wave properties. Heisenberg showed the validity of the uncertainty principle which states that the uncertainty in momentum multiplied by the uncertainty in the position of an electron must be equal to or greater than $h/4\pi$. This means that the momentum and position of the electron cannot be precisely measured simultaneously. As a result of the uncertainty principle, only the regions in space in an atom where an electron has a high probability of being found can be identified. Schrödinger showed that only certain wave functions, ψ, could be associated with the electron's matter-wave. Each of the wave functions designates an allowable energy for the electron in the atom. The wave functions for each of the allowed energy states are orbitals. Three quantum numbers are used to describe the energies, locations, and shapes of orbitals. A fourth quantum number is used to describe the orientation of an electron.

The principal quantum number, n, has all integer values from 1 to ∞, and describes the energy levels of electrons. Electrons with the same principal quantum number are in the same electron shell. The azimuthal or angular momentum quantum number, l, has all integer values from 0 to n – 1. This number describes the subshells within shells. The four lowest energy subshells are the *s*, *p*, *d*, and *f*. The magnetic quantum number, m_l, gives us the number of orbitals within a subshell, and can have all integer between $+l$ and $-l$. The spin quantum number, m_s, describes the orientation of the electron with respect to its apparent spin. The Pauli exclusion principle states that no two electrons in an atom can have the same set of four quantum numbers. This leads to the fact that only two electrons can populate an orbital and these two electrons must have opposite spins.

The electron configuration is a description of the distribution of electrons in energy levels, subshells, and orbitals. The aufbau principle states that electrons fill the lowest energy orbitals before they fill higher energy orbitals. As electrons fill orbitals they obey Hund's, rule which states that the most stable arrangement of electrons in a subshell is the one that has the greatest number of unpaired electrons with parallel spins.

Test Yourself

1. List five types of electromagnetic radiation in order of increasing frequency.

2. Electromagnetic waves *X* and *Y* have wavelengths of 5.39×10^{-8} m and 10.5 nm, respectively. Compare the frequencies of waves *X* and *Y*.

3. Heating sodium releases light with many different wavelengths. One of these has a wavelength of 589 nm. Calculate the energy associated, ΔE, with one quantum of this light.

4. a. What is the photoelectric effect? b. What was its importance in the development of the modern concept of the atom?

5. Use the Balmer equation to calculate the wavelength in nm of light emitted by hydrogen when $n = 3$.

6. An electron in a H atom moves from $n = 2$ to $n = 1$. Calculate the frequency in Hz of the photon released as a result of this electron transition.

7. If sufficient energy is added to an atom, it ionizes and produces a cation. When an electron is removed from a gaseous H atom, a gaseous H ion results, H$^+$(*g*).

$$H(g) \rightarrow H^+(g) + e^-$$

The minimum energy needed to remove an electron from a neutral gaseous atom is called the first ionization energy. a. Calculate the first ionization energy for a H atom, assuming that the electron is initially in the ground state. b. What is the first ionization energy in MJ for 1.000 mol H atoms? The Rydberg constant is 2.179×10^{-18} J.

8. If an electron has a velocity of 5.0×10^5 m/s, what is its wavelength in m?

9. If the uncertainty in measuring the position of an electron moving along the *X* axis is 1.0×10^{-3} nm, what is the uncertainty in its velocity? For this problem use the diameter of a H atom which is 0.08 nm. The mass of the electron is 9.11×10^{-31} kg, *h* is 6.626×10^{-34} J s (kg m^2 s/s^2), and π is 3.14.

10. a. Write the complete and abbreviated electron configurations, showing the order in which the electrons fill, for radium, Ra. b. Write the set of four quantum numbers for the outermost electrons of Ra.

11. Gamma rays, γ, are penetrating waves released by unstable nuclei. What is the wavelength of a γ ray in pm that has a frequency of 7.3×10^{22} Hz?

12. What is the ΔE in J of an atom that releases a photon with a wavelength of 3.12×10^{-7} m?

13. An electromagnetic radiation has a frequency of 3.33×10^{13} Hz. What is the energy in eV of one mole of photons with this frequency? One electron volt, eV, is 1.602×10^{-19} J.

14. What is the kinetic energy of an electron ejected from the surface of Pt when a beam of photons with a wavelength of 200 nm? The binding energy of a Pt electron is 8.01×10^{-19} J.

15. One of the lines in the hydrogen spectrum has a wavelength of 4.102×10^{-5} cm and is the result of an electron falling back to the second energy level. Calculate the value of n that corresponds to this wavelength.

16. List three reasons the Bohr model of the atom is not the one used today.

17. What is the importance of the Schrödinger equation?

18. What do each of the following quantum numbers describe? a. principal b. azimuthal c. magnetic d. spin

19. Write the set of four quantum numbers for the two highest energy electrons in Ca.

20. What is the meaning of the quantum numbers: $n = 3$ $l = 1$?

21. What quantum numbers describe the $4d$ subshell?

22. What is the maximum number of electrons in an atom that could be found in the allowed orbitals with $n = 3$?

23. What is wrong with the statement that all electrons spin about an imaginary axis?

24. What is the difference in the shape and size of the $3s$ and $4s$ subshells?

25. When an atom is ionized, the highest-energy electron leaves first. Write the electron configuration for Ag^+.

26. Calculate the longest possible wavelength in the visible spectrum of hydrogen.

27. The laser used to read information from a compact disk has a wavelength of 780 nm. What is the energy associated with one photon of this radiation?

28. Calculate the number of photons with a wavelength of 10.0 μm needed to produce 1.0 kJ.

29. The minimum energy needed to overcome the binding energy between an electron and surface of silver is 7.5×10^{-19} J. What minimum wavelength in nm would be needed to eject electrons with an energy of 1.0×10^{-18} J/e^-?

30. Which atom is first to have a completely filled $4f$, $5d$, and $6s$? Write its electron configuration.

31. Calculate the wavelength, λ, and frequency, ν, of the wave associated with a proton ($m_{p^+} = 1.673 \times 10^{-24}$ g) with a velocity of 1.5×10^6 m/s.

32. How many unpaired electrons are in Cr and Cr^{3+}?

33. In the ground state of tin, how many electrons have $l = 1$ as one of their quantum numbers?

34. a. What is the physical significance of ψ?
    b. What is the physical significance of ψ^2?

35. Calculate the wavelength in nm of an electromagnetic wave that has a frequency of 7.00×10^{14} Hz.

    a. 4.28×10^{-7} nm
    b. 2.33×10^6 nm
    c. 428 nm
    d. 0.428
    e. none of these

36. One of the principal wavelengths associated with the spectrum of Cd is at 2288 Å. If one angstrom, Å, is equivalent to 1×10^{-10} m, calculate the energy associated with one quantum of this light.

    a. 8.682×10^{-19} J
    b. 8.682×10^{-29} J
    c. 8.682×10^{-20} J
    d. 1.516×10^{-30} J
    e. none of these

37. Which of the following is not a postulate of the Bohr model of the atom?

   a. The electron in the hydrogen atom can only exist in certain fixed quantized energy states.

   b. An electron in a hydrogen atom in any one of its stationary states does not emit electromagnetic radiation.

   c. When hydrogen's electron absorbs or emits electromagnetic radiation of frequency ν, its change in the energy, ΔE, equals $h\nu$.

   d. An electron in a stationary state of a hydrogen atom moves in a circular path called an orbit and obeys the laws of classical physics.

   e. all of these are postulates of the Bohr model of the atom

38. Calculate the frequency in Hz of light released when an electron in a H atom drops from $n = 6$ to $n = 3$.

   a. 1.095×10^{-6} Hz
   b. 1.815×10^{-19} Hz
   c. 3.196×10^{15} Hz
   d. 2.740×10^{14} Hz
   e. none of these

39. Calculate the first ionization energy for a H atom in which its electron is initially in $n = 4$.

   a. 2.420×10^{-19} J
   b. 1.362×10^{-19} J
   c. 2.178×10^{-18} J
   d. 8.712×10^{-20} J
   e. none of these

40. Which of the following equations expresses de Broglie's hypothesis?

   a. $\nu = R_H/h(1/n_f^2 - 1/n_i^2)$
   b. $\lambda = h/(mv)$
   c. $\nu = c/\lambda$
   d. $\Delta E = hc/\lambda$
   e. none of these

Check Yourself

1. Radio waves < microwaves < infrared < visible light < ultraviolet < x-rays < gamma rays **(Electromagnetic radiation)**

2. $\nu = c/\lambda = (2.998 \times 10^8 \text{ m/s})/5.39 \times 10^{-8} \text{ m} = 5.56 \times 10^{15}$ Hz $= 0.556 \times 10^{16}$ Hz

   $10.5 \text{ nm} \times (1 \text{ m}/1 \times 10^9 \text{ nm}) = 1.05 \times 10^{-8}$ m

   $\nu = c/\lambda = (2.998 \times 10^8 \text{ m/s})/1.05 \times 10^{-8} \text{ m} = 2.86 \times 10^{16}$ Hz

   Because the wavelength of X is longer than that of Y, the frequency of X, 5.56×10^{15} Hz, is lower than that of Y, 2.86×10^{16} Hz. **(Electromagnetic radiation)**

3. $589 \text{ nm} \times (1 \text{ m}/1 \times 10^9 \text{ nm}) = 5.89 \times 10^{-7}$ m

   $\Delta E = hc/\lambda = (6.626 \times 10^{-34} \text{ J s} \times 2.998 \times 10^8 \text{ m/s})/5.89 \times 10^{-7} \text{ m} = 3.38 \times 10^{-19}$ J **(Quantum theory)**

4. a. Mathematically, the photoelectric effect is expressed as follows.

   $\frac{1}{2}mv^2 = h\nu - E_{binding}$

   This equation states that the kinetic energy of electrons emitted by photons from surfaces equal the energy of the photon, $h\nu$, minus the binding energy of the electron. The binding energy is the amount of energy needed to break the electron free from the atom to which it is bonded.

   b. Einstein showed that quantum theory developed by Planck could be used to explain phenomena such as the photoelectric effect. The explanation of the photoelectric effect helped validate the quantum theory. **(Photoelectric effect)**

5. $1/\lambda = 1.097 \times 10^7 \text{ m}^{-1} (1/2^2 - 1/n^2) = 1.097 \times 10^7 \text{ m}^{-1} (1/4 - 1/3^2) = 1.524 \times 10^6 \text{ m}^{-1}$

   $= 1/1.524 \times 10^6 \text{ m}^{-1} = 6.563 \times 10^{-7} \text{ m} \times (1 \times 10^9 \text{ nm/m}) = 656.3$ nm **(Bohr atom)**

6. $\nu = R_H/h(1/n_f^2 - 1/n_i^2) = -2.179 \times 10^{-18}$ J$/6.626 \times 10^{-34}$ J s$(1/1^2 - 1/2^2) = -2.466 \times 10^{15}$ Hz
   (Bohr atom)

7. a. $\Delta E = R_H(1/n_f^2 - 1/n_i^2)$

   $\Delta E = -2.179 \times 10^{-18}$ J $(1/\infty^2 - 1/1^2) = 2.179 \times 10^{-18}$ J $(0 - 1) = 2.179 \times 10^{-18}$ J

   b. 2.179×10^{-18} J/atom $\times 6.022 \times 10^{23}$ atoms/mol $\times$ 1 MJ$/1 \times 10^6$ J = 1.312 MJ/mol   **(Bohr atom)**

8. $\lambda = h/(mv) = 6.6 \times 10^{-34}$ J s$/(9.1 \times 10^{-31}$ kg $\times 5.0 \times 10^5$ m/s$) = 1.5 \times 10^{-8}$ m **(de Broglie hypothesis)**

9. $\Delta x \cdot \Delta p \geq h/4\pi$

   $\Delta x \cdot m \Delta v \geq h/4\pi$

   $\Delta v = \dfrac{h}{m \Delta x\, 4\pi}$

   $\Delta v = 6.626 \times 10^{-34}$ J s$/(9.11 \times 10^{-31}$ kg $\times 1 \times 10^{-12}$ m $\times 4 \times 3.14) = 5.8 \times 10^7$ m/s
   (Uncertainty principle)

10. a. Ra $1s^2\ 2s^2\ 2p^6\ 3s^2\ 3p^6\ 4s^2\ 3d^{10}\ 4p^6\ 5s^2\ 4d^{10}\ 5p^6\ 6s^2\ 4f^{14}\ 5d^{10}\ 6p^6\ 7s^2$

    b. Because the outermost electrons are in the 7s subshell, the first three quantum numbers are $n = 7$, $l = 0$, and $m_l = 0$. Because these electrons must have apparent spins in opposite directions, their m_s values are $\pm\frac{1}{2}$. **(Quantum numbers)**

11. $\lambda = c/\nu = (2.998 \times 10^8$ m/s$)/(7.3 \times 10^{22}$ Hz $\times (10^{12}$ pm/m$)) = 4.1 \times 10^{-3}$ pm **(Quantum theory)**

12. $\Delta E = hc/\lambda = (6.626 \times 10^{-34}$ J s $\times 2.998 \times 10^8$ m/s$)/3.12 \times 10^{-7}$ m $= 6.37 \times 10^{-19}$ J   **(Quantum theory)**

13. $E = h\nu = 6.626 \times 10^{-34}$ J s $\times 3.33 \times 10^{13}$ Hz $\times (1$ eV$/1.602 \times 10^{-19}$ J$) = 0.138$ eV   **(Quantum theory)**

14. 200 nm $\times (1$ m$/1 \times 10^9)$ nm $= 2.00 \times 10^{-7}$ m

    $\nu = c/\lambda = (2.998 \times 10^8$ m/s$)/2.00 \times 10^{-7}$ m $= 1.50 \times 10^{15}$ Hz

    $E_{kinetic} = h\nu - E_{binding} = (6.626 \times 10^{-34}$ J s $\times 1.50 \times 10^{15}$ Hz$) - 8.01 \times 10^{-19}$ J $= 1.92 \times 10^{-19}$ J
    (Photoelectric effect)

15. 410.2 nm $\times (1$ m$/1 \times 10^9$ nm$) = 4.102 \times 10^{-7}$ m

    $hc/\lambda = R_H(1/n_f 2 - 1/n_i 2)$

    $(6.626 \times 10^{-34}$ J s $\times 2.998 \times 10^8$ m/s$)/4.102 \times 10^{-7}$ m $= -2.179 \times 10^{-18}$ J $(1/2^2 - 1/n^2)$

    $n = 6$ **(Bohr atom)**

16. 1. The Bohr model could only explain one-electron systems.

    2. The Bohr model could not explain the intensities of the lines in the H spectrum.

    3. The Bohr model did not incorporate the uncertainty principle and the wave nature of the electron.
    (Bohr atom)

17. The solution of the Schrödinger equation gives a wave function, ψ. The wave function is often called an orbital. This equation incorporated the wave and quantum nature of the electron to predict the orbitals in an atom. **(Wave equation)**

18. a. The principal quantum numbers give the energy levels (size) for electron orbitals.

    b. The azimuthal quantum numbers give the sublevels (shapes) for electron orbitals.

    c. The magnetic quantum numbers give the orientations of orbitals; e.g., p_x, p_y, and p_z.

    d. The spin quantum numbers give the apparent spins of electrons. **(Quantum numbers)**

19. $n = 4, l = 0, m_l = 0, m_s = +½$
    $n = 4, l = 0, m_l = 0, m_s = -½$ **(Quantum numbers)**

20. When $n = 3$, this describes the third energy level. When $l = 1$, the sublevel is a p. Hence, this set of quantum numbers describes the $3p$ sublevel. **(Quantum numbers)**

21. $n = 4, l = 2$ **(Quantum numbers)**

22. When $n = 3$, the values of l can be 0 (s), 1 (p), or 2 (d). The s, p, and d hold a maximum of 2, 6, and 10 electrons, respectively. Therefore, a total of 18 electrons in an atom can have a n value of 3. **(Quantum numbers)**

23. The uncertainty principle precludes the possibility of seeing electrons; hence, we can never know if electrons are spinning. The best statement is that the properties of electrons indicate that electrons appear to be spinning. **(Uncertainty principle)**

24. Because the n value is the only difference, these subshells only differ in their size. The $3s$ subshell is smaller and on an average closer to the nucleus than the $4s$ subshell. **(Electron configuration)**

25. Ag^+ [Kr] $4d^{10} 5s^0$ **(Electron configuration)**

26. The visible spectrum of hydrogen arises from electron transitions to the second energy level. To produce the longest wavelength, it must have the lowest frequency, which means the smallest ΔE. Therefore, the transition from 3 to 2 produces the longest wavelength.
    $hc/\lambda = R_H(1/n_f^2 - 1/n_i^2)$
    $(6.626 \times 10^{-34} \text{ J s} \times 2.998 \times 10^8 \text{ m/s})/\lambda = -2.179 \times 10^{-18} \text{ J } (1/2^2 - 1/3^2)$
    $\lambda = 6.567 \times 10^{-7} \text{ m} = 656.7 \text{ nm}$ **(Bohr atom)**

27. $780 \text{ nm} \times (1 \text{ m}/10^9 \text{ nm}) = 7.80 \times 10^{-7} \text{ m}$
    $\Delta E = hc/\lambda = (6.626 \times 10^{-34} \text{ J s} \times 2.998 \times 10^8 \text{ m/s})/7.80 \times 10^{-7} \text{ m} = 2.55 \times 10^{-19} \text{ J}$ **(Quantum theory)**

28. $10.0 \text{ μm} \times (1 \text{ m}/10^6 \text{ nm}) = 1.00 \times 10^{-5} \text{ m}$
    $\Delta E = hc/\lambda = (6.626 \times 10^{-34} \text{ J s} \times 2.998 \times 10^8 \text{ m/s})/1.00 \times 10^{-5} \text{ m} = 1.99 \times 10^{-20} \text{ J/photon}$
    $1 \text{ kJ} \times (1000 \text{ J/kJ}) \times (1 \text{ photon}/1.99 \times 10^{-20} \text{ J}) = 5.03 \times 10^{22}$ photons **(Quantum theory)**

29. $hc/\lambda = E_{binding} + E_{kinetic}$
    $(6.626 \times 10^{-34} \text{ J s} \times 2.998 \times 10^8 \text{ m/s})/\lambda = 7.5 \times 10^{-19} \text{ J} - 1.0 \times 10^{-18} \text{ J}$
    $\lambda = 1.14 \times 10^{-7} \text{ m}$ **(Photoelectric effect)**

30. Hg The configuration of Hg is [Xe] $4f^{14} 5d^{10} 6s^2$. **(Electron configuration)**

31. $\lambda = h/(mv) = 6.626 \times 10^{-34} \text{ J s}/(1.673 \times 10^{-27} \text{ kg} \times (1 \text{ kg}/1000 \text{ g}) \times 1.5 \times 10^6 \text{ m/s}) = 2.64 \times 10^{-13} \text{ m}$
    $\nu = c/\lambda = (2.998 \times 10^8 \text{ m/s})/2.64 \times 10^{-13} \text{ m} = 1.14 \times 10^{21} \text{ Hz}$ **(Quantum theory)**

32. Cr [Ar] $3d^5 4s^1$ - six unpaired electrons
    Cr^{3+} [Ar] $3d^3 4s^0$ - three unpaired electrons **(Electron configuration)**

33. Sn [Kr] $4d^{10} 5s^2 5p^2$
    When $l = 1$ the sublevel is the p. Tin atoms have the $2p^6$, $3p^6$, $4p^6$, and $5p^2$, which gives a total of 20 electrons with $l = 1$. **(Electron configuration)**

34. a. The solution of the Schrödinger equation gives a wave function, ψ, which describes an orbital in an atom.

    b. The value of ψ^2 is proportional to the probability of finding an electron in a small volume of space outside the nucleus. **(Wave equation)**

35. c. $\lambda = c/\nu = 2.998 \times 10^8$ m/s/7.00×10^{14} Hz $\times (10^9$ nm/m$) = 428$ nm **(Electromagnetic waves)**

36. a. 2288 Å $\times 1$ m/10^{10} Å $= 2.288 \times 10^{-7}$ m

    $\Delta E = hc/\lambda = (6.626 \times 10^{-34}$ J s $\times 2.998 \times 10^8$ m/s$)/2.288 \times 10^{-7}$ m $= 8.682 \times 10^{-19}$ J   **(Quantum theory)**

37. d. **(Bohr atom)**

38. d. $\nu = (R_H/h)(1/n_f^2 - 1/n_i^2)$

    $\nu = -2.179 \times 10^{-18}$ J/6.626×10^{-34} J s $(1/3^2 - 1/6^2) = -2.740 \times 10^{14}$ Hz **(Bohr atom)**

39. b. $\Delta E = R_H(1/n_f^2 - 1/n_i^2)$

    $\Delta E = -2.179 \times 10^{-18}$ J $(\dfrac{1}{\infty^2} - \dfrac{1}{4^2}) = 1.362 \times 10^{-19}$ J **(Bohr atom)**

40. b. $\lambda = h/(mv)$ **(de Broglie hypothesis)**

Grade Yourself

Circle the numbers of the questions you missed, then fill in the total incorrect for each topic. If you answered more than three questions incorrectly, you need to focus on that topic. (If a topic has less than three questions and you had at least one wrong, we suggest you study that topic also. Read your textbook, a review book, or ask your teacher for help.)

Subject: The Electronic Structure of Atoms

Topic	Question Numbers	Number Incorrect
Electromagnetic radiation	1, 2	
Quantum theory	3, 31, 36	
Photoelectric effect	4, 14, 29	
Bohr atom	5, 6, 7, 15, 16, 26, 37, 38, 39	
de Broglie hypothesis	8, 40	
Uncertainty principle	9, 23	
Quantum numbers	10, 18, 19, 20, 21, 22	
Quantum theory	11, 12, 13, 27, 28	
Wave equation	17, 34	
Electron configuration	24, 25, 30, 32, 33	
Electromagnetic waves	35	

Periodic Properties of Atoms and Elements

Brief Yourself

Meyer and Mendeleev developed the forerunner to our modern periodic table. They saw that if the elements are arranged according to their atomic masses, then patterns of recurring properties can be identified. Both Meyer and Mendeleev placed elements with similar properties in vertical columns called chemical groups.

The representative elements, transition elements, and inner transition elements are the three principal groups of elements. The outer-level s and p subshells fill in the representative elements. The d subshell fills in the transition elements and the f subshells fill in the inner transition elements.

The periodic law states that if the elements are arranged according to their atomic numbers then their properties recur periodically. Some of the periodic properties of elements are ionization energy, electron affinity, atomic size, and electronegativity. The first ionization energy of an atom is the minimum energy needed to remove the most loosely held electron from a gaseous atom. From left to right across a period the first ionization energies show an irregular increase, and within a group, from top to bottom, the first ionization energies decrease.

The electron affinity of an atom is the enthalpy change, ΔH, when an electron is added to a gaseous atom. Atoms that accept electrons have larger negative electron affinities than those that do not readily accept electrons. Nonmetals tend to have larger negative first electron affinities than metals.

Two measures of the sizes of atoms are the covalent and ionic radii. Proceeding across a period from left to right the size of atoms decreases, and going from top to bottom in a chemical group the size of atoms increases. The ionic radius of a metal cation is smaller than the size of the atom because the outer level electrons are removed. However, the ionic radius of a nonmetallic anion is larger than the size of the atom because of increased repulsions of the additional electron(s) added to the outer level.

Electronegativity is a measure of the power of atoms to attract electrons in chemical bonds. The Pauling electronegativity scale ranges from 0 to 4, with F being the most electronegative atom. Within a period, electronegativity values increase (excluding the noble gases), and within a group the values decrease.

The magnetic properties of atoms and ions result from their electron configuration. Elements are either classified as being diamagnetic or paramagnetic. A diamagnetic species is weakly repelled by magnetic fields because its electrons are paired. A paramagnetic species is attracted by magnetic fields because it has one or more unpaired electrons.

Test Yourself

1. In what chemical group does each of the following belong? a. sulfur b. neon c. calcium

2. What is the name of each of the following chemical groups? a. IA b. B group c. VIIA

3. What are the representative elements?

4. Explain why most periodic tables place H in group IA (1) when it shares no common chemical and physical properties with the alkali metals?

5. a. Write an equation that shows what happens when energy equal to the first ionization energy is added to a potassium atom. b. Write the electron configurations for potassium and the potassium ion.

6. Which of the following elements has the highest and lowest first ionization energies? S, Cl, Se, Br

7. Which of the following elements has the highest third ionization energy? Na, Mg, K, Ca

8. List three factors that determine the magnitude of the first ionization energy of an atom.

9. Explain why the first ionization energy of B is less than that of Be?

10. What trend in first ionization energies is found in the alkali metals? Explain this trend.

11. A relatively small increase in ionization energies is observed going from Sc to Zn when compared to the increase across the representative elements in this period. How can this difference be explained?

12. a. Write a statement that describes the second ionization energy of an atom. b. If the second ionization energy for Na is 4.56 MJ/mol, write an equation that shows this change.

13. a. Write the equation that shows the electron affinity for fluorine. b. Explain why fluorine has a high negative electron affinity value.

14. What group of elements have the highest positive electron affinities? Explain why their values are so high.

15. What are the trends in atomic sizes (atomic radii) across a period and within a group?

16. The changes in the properties across the transition elements are much less than those of the representative elements. a. Explain why the decrease in the atomic size of transition elements is smaller than that for the representative elements. b. What trends in atomic sizes are expected across the lanthanide elements?

17. Explain why the radius of an Al atom ($Z = 13$) is larger than that of a Ga atom ($Z = 31$).

18. a. Write a definition of electronegativity. b. What type of atoms have high electronegativities and what type of atoms have low electronegativities? Explain.

19. What is the difference between a diamagnetic and paramagnetic chemical species? Give an example of each.

20. Which of the following are paramagnetic species? a. Ca b. I^- c. Ag^+ d. Fe^{3+} e. Ni^{2+}

21. Predict the physical state, reactivity, electron configuration, and first ionization energy of the yet-to-be discovered element 118.

22. For each of the following properties, compare element 81, thallium (Tl) to element 34, selenium (Se): a. boiling and melting points b. electrical conductivity c. malleability d. first ionization energy. Explain your answer.

23. For each of the following properties, compare element 31 to element 16 a. boiling and melting points b. thermal conductivity c. first ionization energy d. atomic size

24. a. Write equations for the first three ionizations of Li. b. What is the electron configuration for Li^{3+}?

25. Arrange the following elements in order of their first ionization energies: In, Sn, Ga, Ge.

26. The first, second, and third ionization energies of Al are 0.578, 1.82, and 2.74 MJ/mol, respectively. Calculate the energy needed to remove three electrons from Al(g).

$$Al(g) \rightarrow 3e^- + Al^{3+}(g)$$

27. Explain why the sign of an electron affinity is either positive or negative.

28. Arrange the following in order of increasing radius: Br, Br^-, and Br^+.

29. How is electronegativity different from electron affinity?

30. For each of the following, compare the properties of sodium to potassium. a. first ionization energies, b. electron affinities, c. electro-negativities, d. magnetic properties

31. Which period 2 element has the following properties: a. lowest electronegativity, b. largest atomic radius, c. lowest boiling point, d. highest second ionization energy, e. highest third ionization energy, f. most negative electron affinity?

32. The first ionization energies of F, Cl, and I are 1.68, 1.25, and 1.01 MJ/mol, respectively. Predict the value for the first ionization energy of bromine.

33. How many Li atoms could be ionized from the energy released when 1.00 mol of electrons are added to 1.00 mol of gaseous Cl atoms? The first ionization energy of Li is 0.52 MJ/mol. The electron affinity of Cl is –349 kJ/mol.

34. The electron affinity of O(g) is –142 kJ/mol, but the electron affinity for $O^-(g)$ is +780 kJ/mol. Explain why O(g) favorably accepts an electron, but $O^-(g)$ does not, even though a noble gas configuration results.

35. Which of the following has the highest first ionization energy?
    a. C
    b. N
    c. Si
    d. P
    e. Ge

36. Which requires the greatest input of energy?
    a. first ionization energy of Na
    b. second ionization energy of Na
    c. third ionization energy of Na
    d. fourth ionization energy of Na

37. Predict which of the following has the most positive electron affinity.
    a. Br
    b. Sr
    c. Kr
    d. S

38. Which of the following is the largest atom?
    a. Rb
    b. Sr
    c. Cs
    d. Ba

39. Which of the following is the smallest?
    a. O^{2-}
    b. F^-
    c. Na^+
    d. Mg^{2+}

40. Which of the following has the highest electronegativity?
    a. Rb
    b. Sr
    c. Cs
    d. Ba

Check Yourself

1. a. chalcogens (VIA, 16), b. noble gases (VIIIA, 18), c. alkaline earth metals (IIA, 2) **(Periodic table)**

2. a. alkali metals, b. transition metals (transition elements), c. halogens **(Periodic table)**

3. The representative elements are those in groups with A designations (IA, IIA, IIIA to VIIIA). They include all of the elements except for the transition metals. **(Periodic table)**

4. All members of group IA (1) have ns^1 for their valence electron configurations. This periodic grouping places all elements with one valence electron in the same chemical group and neglects the differences in properties. Hydrogen is a nonmetal and belongs in its own group. **(Periodic properties)**

5. a. $K \rightarrow K^+ + e^-$ $\Delta H = I_1$
   b. K [Ar] $4s^1$, K^+ [Ar] $4s^0$ **(Ionization energy)**

6. First ionization energies generally increase across a period and decrease within a chemical group. Thus, Cl has the highest first ionization energy because it is farther to the right in the period and a lower-molecular mass member of group 17. Selenium, Se, has the lowest first ionization energy of this group because it is closest to the left side of the periodic table and is the higher-molecular mass member of group 16. **(Ionization energy)**

7. The atom that an electron removed from the noble gas configuration requires the greatest energy. The third ionization energy removes the most loosely-held electron from a dipositive cation. Hence, the ionization energies of Mg and Ca, the group 2 members, are higher than Na and K, respectively, because Mg^{2+} and Ca^{2+} have noble gas configurations. Finally, Mg^{2+} should have a higher third ionization energy than Ca^{2+} because the outermost electrons are in a lower energy shell than those of Ca^{2+}. **(Ionization energy)**

8. Effective nuclear charge, inner electron shielding, and distance from the nucleus. **(Ionization energy)**

9. The valence electron configurations for Be and B are $2s^2$ and $2s^2 2p^1$, respectively. Usually, an electron in a $2p$ orbital tends to be farther from the nucleus than an electron in a $2s$ orbital. The farther an electron is from the nucleus, the weaker the force of attraction by the nucleus. Additionally, the $2s$ electrons shield nuclear charge from the $2p$ electron. Hence, the first ionization energy decreases. **(Ionization energy)**

10. Going from Li to Fr, the ionization energies decrease. The outermost electron in Li is in the $2s$ and the outermost in Na is the $3s$. The average distance of an electron in the $3s$ is farther from the nucleus than one in the $2s$. The farther an electron is from the nucleus, the easier it is to remove because it is in a higher-energy state. **(Ionization energy)**

11. Electrons in lower-energy subshells shield nuclear charge from those in higher-energy subshells. Thus, electrons that enter the $3d$ subshell shield nuclear charge from the electrons in the $4s$ subshell and partially counterbalance the increase in nuclear charge going from Sc ($Z = 21$) to Zn ($Z = 30$). Hence, only a small increase in ionization energy is observed. This shielding effect is not found in the representative elements. **(Ionization energy)**

12. a. The second ionization energy, I_2, is the minimum energy required to remove the most loosely-held electron from a monopositive gaseous cation, $A^+(g)$.
    b. $Na^+(g) \rightarrow Na^{2+}(g) + e^-$ $I_2 = 4.56$ MJ/mol **(Ionization energy)**

13. a. $F(g)(2s^2 2p^5) + e^- \rightarrow F^-(g)(2s^2 2p^6)$ $\Delta H_{\text{Electron affinity}} = -322$ kJ/mol

b. When a halogen atom such as F accepts an electron, it obtains the stable noble gas electron configuration and thus releases energy as it goes to a lower energy state. The large negative electron affinity of F is the result of its high effective nuclear charge. **(Electron affinity)**

14. The alkaline earth metals, group 2, have the most positive electron affinities. The large positive first electron affinities of this group shows that they tend not to form anions because it adds an electron to the higher-energy p subshell. The alkali metals have less positive electron affinities because there is space in the s subshell to accept the electron, resulting in a filled orbital. **(Electron affinity)**

15. Atoms decrease in size going from left to right across a period, and increase in size going from top to bottom in a chemical group. **(Atomic size)**

16. a. The added inner level d electrons shield nuclear charge from the outer level s electrons.

    b. A similar gradual decrease in atomic size occurs across the inner transition elements in which the $4f$ subshell fills. **(Atomic size)**

17. This apparent anomaly is explained in terms of the incomplete shielding of the d^{10} electrons. Due to the more elongated shape of d orbitals, electrons in the d subshell do not shield as effectively as s or p electrons. Incomplete shielding results in a higher effective nuclear charge, which means a stronger force of attraction on the $4p$ electrons. **(Atomic size)**

18. a. Electronegativity is the power that an atom has to attract electrons in a chemical bond.

    b. More compact atoms (nonmetals), those with high effective nuclear charges, tend to have higher electronegativities. More diffuse atoms (metals), those with lower effective nuclear charges, tend to have lower electronegativities. **(Electronegativity)**

19. A diamagnetic species has all paired electrons (↑↓) while a paramagnetic species has one or more unpaired electrons (↑). Noble gases ($ns^2 np^6$) are diamagnetic because they have a filled outer s and p orbitals, which means their electrons are paired. Halogens ($ns^2 np^5$) are paramagnetic because they have one unpaired electron in their p orbitals. **(Magnetic properties)**

20. The paramagnetic species are Fe^{3+} and Ni^{2+}. The electron configuration of Fe^{3+} is [Ar] $3d^5$, which means that it has five unpaired electrons. The electron configuration of Ni^{2+} is [Ar] $3d^8$, which means that it has two unpaired electrons. All of the others have paired electrons. **(Magnetic properties)**

21. Element 118 will belong to the noble gases in period 7. All of the other elements in group 18 are gases at 25°C; hence, it should be a gas. It will be the least reactive element in period 7 and the most reactive noble gas. It is expected to have an outer-level electron configuration of $7s^2 7p^6$. Because the first ionization energy decreases within a group and increases in a period, Element 118 should have the lowest first ionization energy of the noble gases and the highest ionization energy in period 7. **(Predicting properties)**

22. Thallium is the heaviest member of group 13 (III A); therefore, Tl is a heavy metal. Selenium is the third member of the chalcogens, group 16 (VI A); hence, it has the properties of a nonmetal.

    a. Most often the boiling and melting points of metals are higher than those of nonmetals. Therefore, it is a prediction that the boiling and melting points of Tl are higher than those of Se.

    b. The electrical conductivity of metals is greater than that of nonmetals; thus, Tl is predicted to be a better conductor than Se.

    c. Because Tl is a metal it is malleable. Se is a nonmetal and thus not malleable.

    d. From its placement on the periodic table, Tl is a large metal atom and Se is a much smaller nonmetal atom. Hence, the first ionization energy of Se is higher than that of Tl. **(Predicting properties)**

23. a. Ga has higher melting and boiling points than S. b. Ga is a better thermal conductor than S. c. Ga has a lower first ionization energy than S. d. Ga is composed of larger atoms than S. (**Predicting properties**)

24. a. Li → Li$^+$ + e$^-$ I_1,   Li$^+$ → Li^{2+} + e$^-$ I_2 Li^{2+} → Li^{3+} + e$^-$ I_3
    b. Li^{3+} has no electrons. (**Ionization energy**)

25. Ge > Sn > Ga > In (**Ionization energy**)

26. Use Hess' law by adding the first three ionization energies to obtain the overall energy needed to remove the first three electrons of Al.
    ΔE = 0.578 MJ/mol + 1.92 MJ/mol + 2.74 MJ/mol = 5.14 MJ/mol. (**Ionization energy**)

27. When electrons are added to atoms either energy is released or absorbed. If energy is released the sign is negative, and if it is absorbed it is positive. (**Electron affinity**)

28. Br$^-$ > Br > Br$^+$ The anion has one more electron than proton; thus, the effective nuclear charge is decreased. The cation has one more proton than electron; thus, the effective nuclear charge is increased. (**Atomic size**)

29. Electronegativity is the capacity of atoms to attract electrons in chemical bonds. Electronegativity is principally applied to explain bonding phenomena that involve two atoms attracting overlapping electrons. Electron affinity gives the energy transfer when an electron is added to a neutral gaseous atom. It only involves an isolated atom. (**Electronegativity**)

30. a. Na > K, b. Na > K, c. Na > K, d. both paramagnetic (**Magnetic properties**)

31. a. Li, b. Li, c. Ne, d. Li, e. Be, f. F (**Predicting properties**)

32. Taking the average of the first ionization energies of Cl and I gives 1.13 MJ/mol for the value of Br. The actual value for Br is 1.14 MJ/mol. (**Predicting properties**)

33. 349 kJ × (1 MJ/1000 kJ) × (1 mol Li/0.52 MJ) × (6.02× 10^{23} atoms/mol) = 4.0 × 10^{23} atoms Li   (**Ionization energy**)

34. In the gas phase, the electron-electron repulsions in O$^-$ are greater than the stabilization that results from obtaining a noble gas configuration. Also negative ions do not attract electrons as readily as neutral atoms. Note, the oxide, O^{2-}, is a common ion in ionic solids, which have additional attractive forces that give produce a more stable ion. (**Electron affinity**)

35. b. Nitrogen is farther to the right and higher up on the periodic table than the other elements. (**Ionization energy**)

36. d. The higher-order ionization always requires a greater amount of energy because the electron is being removed from a more positive cation. (**Ionization energy**)

37. b. Alkaline earth metals have the most positive electron affinities of the elements. (**Electron affinity**)

38. c. Rb and Cs are members of group 1, and Sr and Ba are members of group 2. Larger atoms are farther to the left within a period and have higher energy outer-level electrons in a group. Thus, Cs is the largest of this group. (**Atomic size**)

39. d. The ionic radii of nonmetals is greater than their atomic radii. When metal atoms such as Na and Mg lose electrons, their ionic radii decrease. Because all of these ions are isoelectronic to Ne, the cations should be smaller than the anions. Because Mg loses two electrons and Na only loses one electron, the size of Mg^{2+} should be smaller than Na$^+$. (**Atomic size**)

40. b. Elements nearer to the top and farther to the right on the periodic table have higher electronegativities than those that are lower and farther to the left. **(Electronegativity)**

Grade Yourself

Circle the numbers of the questions you missed, then fill in the total incorrect for each topic. If you answered more than three questions incorrectly, you need to focus on that topic. (If a topic has less than three questions and you had at least one wrong, we suggest you study that topic also. Read your textbook, a review book, or ask your teacher for help.)

Subject: Periodic Properties of Atoms and Elements

Topic	Question Numbers	Number Incorrect
Periodic table	1, 2, 3	
Periodic properties	4	
Ionization energy	5, 6, 7, 8, 9, 10, 11, 12, 13, 24, 25, 26, 33, 35, 36	
Electron affinity	14, 27, 34, 37	
Atomic size	15, 16, 17, 28, 38, 39	
Electronegativity	18, 29, 40	
Magnetic properties	19, 20, 30	
Prediction properties	21, 22, 23, 31, 32	

Chemical Bonds– Fundamental Concepts

 ## Brief Yourself

Chemical bonds are the forces of attractions among atoms and ions. Ionic bonds result when one or more electrons transfer between atoms or groups of atoms. The resulting cations and anions attract each other as a result of their opposite charges. The force of attraction between oppositely charged ions is an ionic bond. In the formation of an ionic bond, atoms of representative elements obtain a stable noble gas electron configuration. Few transition elements can obtain a noble gas configuration; however, some obtain a pseudonoble gas configuration.

Ions in an ionic compound are in a specific crystal structure, which is a regular network of alternating cations and anions. Lattice energy is the energy required to dissociate the crystal lattice structure to individual ions in the gas phase. Ionic compounds that have ions with higher charges that are more closely packed tend to have higher lattice energies than those with lower charges that are more widely separated.

Covalent bonds result when one or more of the outer orbitals of one nonmetal atom overlaps with one or more outer orbitals from another nonmetal atom. The covalent bond results from the attraction of the nonmetal nuclei for the overlapping orbitals. The electrons in overlapping orbitals are termed shared or bonded electrons. If nonmetallic atoms with the same electronegativity bond, they form nonpolar covalent bonds in which the atoms share the electrons equally. If atoms with different electronegativities bond, they form polar covalent bonds in which the atoms do not share the electrons equally. Polar diatomic molecules have dipole moments greater than zero, and nonpolar diatomic molecules have zero dipole moments. The dipole moment, μ, of a molecule is calculated from the product of its partial charges and distance that separates them. Dipole moments are measured by placing molecules in electric fields.

Covalent bonds may be characterized by their bond orders, bond distances, and bond energies. Bond order is the number of covalent bonds between two atoms. Bond distance is the average distance between the nuclei that form a covalent bond. Bond energy is the energy needed to cleave a covalent bond. As the bond order increases, the bond distance decreases, and the bond energy increases. Triple bonds are usually shorter and stronger, and single bonds are usually longer and weaker.

Resonance is exhibited in molecules that have more than one Lewis structure that differ only in the placement of the electrons. Each of the Lewis structures written for a molecule that exhibits resonance is called a contributing structure. The average of all of the most important contributing structures best approximates the actual bonding in the molecule.

While most molecules obey the octet rule, many molecules have atoms that do not achieve noble gas configurations. These molecules are generally less stable than similar molecules that have atoms with noble gas configurations.

Test Yourself

1. a. What is a chemical bond? b. Why do chemical bonds form?

2. What are the names of the two bonding theories?

3. a. Write equations that show what happens when Na and Cl combine to produce NaCl(s). b. Show the electron configurations for all species. c. What atoms are the sodium and chloride ions isoelectronic to?

4. What is the formula of the ionic compound that results when oxygen combines with the following metals: a. Ba, b. Al, c. Na?

5. a. How many electrons are transferred when magnesium and fluorine combine? b. What noble gas atoms are the magnesium and fluoride ions isoelectronic to?

6. Iron forms both iron(II) chloride and iron(III) chloride. a. Write the formulas of these two iron chlorides. b. What are the electron configurations of the iron(II) and iron(III) ions?

7. Use the following thermodynamic data to calculate the enthalpy change for the formation of solid lithium fluoride, LiF(s), from Li(s) and $F_2(g)$

$$Li(s) + \tfrac{1}{2}F_2(g) \rightarrow LiF(s)$$

$Li(s) \rightarrow Li(g)$	$\Delta H° = 155$ kJ
$\tfrac{1}{2}F_2(g) \rightarrow F(g)$	$\Delta H° = 75$ kJ
$Li(g) \rightarrow e^- + Li^+(g)$	$\Delta H° = 520$ kJ
$F(g) + e^- \rightarrow F^-(g)$	$\Delta H° = -333$ kJ
$Li^+(g) + F^-(g) \rightarrow LiF(s)$	$\Delta H° = -1012$ kJ

8. What factors are the most important in determining the magnitude of the an ionic solid's lattice energy?

9. Explain why the lattice energy of NaF, 933 kJ/mol, is significantly smaller than that of MgF_2, 2910 kJ/mol.

10. What are the general properties of ionic compounds?

11. What is the bond order for each of the following diatomic molecules: a. Cl_2, b. N_2, c. O_2? Explain.

12. Rank the following from highest to lowest bond energies and longest to shortest bond distances: C–C, C=C, and C≡C.

13. Arrange the following bonds in increasing order, lowest to highest, of bond distance and bond energies: N=N, N≡N, C=C, and C≡C. Explain.

14. Calculate the enthalpy change for the following reaction.

$$H_2(g) + F_2(g) \rightarrow 2HF(g)$$

The bond energies for H–H, F–F, and H–F are 432, 155, and 565 kJ/mol, respectively. State if the reaction is exothermic or endothermic.

15. The actual dipole moment of HCl is 1.07 D. If the charge on a proton is 1.60×10^{-19} C and the bond distance of a HCl molecule is 1.27×10^{-10} m, calculate the percent ionic character of HCl. (1 D = 3.34×10^{-30} C m)

16. Draw the Lewis structure of carbon disulfide.

17. Draw the Lewis structure of potassium phosphide.

18. Draw the Lewis structure of formaldehyde, H_2CO.

19. Draw the Lewis structure of ClF_3.

20. Draw the three principal resonance (contributing) structures for N_2O.

21. Compare the formal charges on the following two resonance (contributing) structures and determine which more closely resembles the actual bonding in hydrogen cyanide.

    H–C≡N: and   H–N≡C:

22. Draw two different Lewis structures for CNBr. Which is the more plausible structure?

23. a. What is resonance? b. How do resonance (contributing) structures differ? c. How are resonance structures separated?

24. Explain why it is incorrect to talk of molecules of solid calcium oxide.

25. The solubility of potassium hydroxide, KOH(s), is 112 g/100 g H_2O at 20°C, whereas the solubility of calcium hydroxide, $Ca(OH)_2$(s), is less than 1 g/100 H_2O at 20°C. Considering their structure and predicted lattice energies, explain the difference in solubility of KOH and $Ca(OH)_2$.

26. Both of the N–O bond distances in the nitrite ion, NO_2^-, are 124 pm. a. Explain why both have the same bond distance. b. What is the bond order of the N–O bond?

27. Arrange the following bonds from least polar to most polar: C–F, Cl–F, N–F, P–F.

28. The dipole moments of HF and HCl are 1.98 D and 1.03 D, respectively. Explain why the dipole moment of HF is significantly higher than that of HCl.

29. The C–O triple bond distance in carbon monoxide is 113 pm. If the percent ionic character of this bond is 2.0%, calculate the experimental value for the dipole moment of CO.

30. Consider the following reaction.
    $$C_2H_2 + Br_2 \rightarrow C_2H_2Br_2$$
    Using the following bond energies, calculate the $\Delta H°$ for this reaction.

Bond	Bond Energy, kJ/mol
C–C	347
C=C	614
C≡C	839
C–Br	276
C–H	413
Br–Br	193
H–Br	363

31. Two different molecules have the molecular formula C_2H_6O. Write the Lewis structures for both of these molecules.

32. Draw three resonance (contributing) structures for the cyanate ion, OCN^-.

33. Draw the Lewis structure of AsF_5.

34. Draw the Lewis structure of $XeOF_4$.

35. Which of the following has the highest lattice energy?
    a. CsF
    b. CsCl
    c. CsBr
    d. CsI
    e. all are the same

36. Which of the following has the longest bond?
    a. ClF
    b. Cl_2
    c. ClBr
    d. ClI
    e. all are the same length

37. Which of the following is in correct order of bond energies.
    a. O–F < C=O < N≡N
    b. O–F > C=O > N≡N
    c. C=O < O–F < N≡N
    d. N≡N < O–F < C=O
    e. none of these

38. Which of the following has the largest dipole moment?

    a. H$_2$
    b. HF
    c. HCl
    d. HBr
    e. HI

39. Which of the following bonds has the greatest percent ionic character?

    a. O–O
    b. O–Cl
    c. O–Br
    d. O–I
    e. none of these

40. The ΔH_f of KF is –563 kJ/mol

    $$K(s) + \tfrac{1}{2}F_2(g) \rightarrow KF(s)$$

    and the energy to vaporize K(s) is +90 kJ/mol.

    $$K(s) \rightarrow K(g)$$

    The first ionization energy of K(g) is +415 kJ/mol. F$_2$(g) requires +155 kJ/mol to dissociate into two F atoms,

    $$F_2(g) \rightarrow 2F(g)$$

    and F$_2$ has a first electron affinity of –322 kJ/mol.

    $$F(g) + e^- \rightarrow F^-(g)$$

    Use this data to calculate the lattice energy of KF.

    a. 824 kJ
    b. 1467 kJ
    c. 871 kJ
    d. 1515 kJ
    e. none of these

Check Yourself

1. a. A chemical bond is the force of attraction that one atom (or ion) has for another atom (or ion). b. The principal "driving force" for atoms to combine and form molecules is the release of energy by atoms as they achieve lower energy states and thus become more stable. **(Chemical bonds)**

2. The two bonding theories are the valence bond and molecular orbital theories. **(Chemical bonds)**

3. Na ($1s^2\ 2s^2\ 2p^6\ 3s^1$) $\rightarrow e^- +$ Na$^+$ ($1s2\ 2s^2\ 2p^6$) **(Isoelectronic to Ne)**
   Cl ($1s^2\ 2s^2\ 2p^6\ 3s^2\ 3p^5$) $+ e^- \rightarrow$ Cl$^-$ ($1s^2\ 2s^2\ 2p^6\ 3s^2\ 3p^6$) (Isoelectronic to Ar) **(Ionic bonds)**

4. a. BaO, b. Al$_2$O$_3$, c. Na$_2$O **(Ionic bonds)**

5. a. Two electrons are transferred from the Mg to two F atoms. b. Both Mg^{2+} and F$^-$ are isoelectronic to Ne. **(Ionic bonds)**

6. a. iron(II) chloride = FeCl$_2$, iron(III) chloride = FeCl$_3$
   b. Fe ([Ar] $3d^6\ 4s^2$) $\rightarrow 2e^- +$ Fe^{2+} ([Ar] $^3d^6$)
   Fe ([Ar] $3d^6\ 4s^2$) $\rightarrow 3e^- +$ Fe^{3+} ([Ar] $^3d^5$) **(Ionic bonds)**

Chemical Bonds—Fundamental Concepts / 69

7. Li(s) → Li(g) $\Delta H° = 155$ kJ

    ½F$_2$(g) → F(g) $\Delta H° = 75$ kJ

    Li(g) → e$^-$ + Li$^+$(g) $\Delta H° = 520$ kJ

    F(g) + e$^-$ → F$^-$(g) $\Delta H° = -333$ kJ

    Li$^+$(g) + F$^-$(g) → LiF(s) $\Delta H° = -1012$ kJ

    Li(s) + ½F$_2$(g) → LiF(s) $\Delta H° = -595$ kJ **(Ionic bonds)**

8. Lattice energies depend mainly on the type of crystal structure, the distance between ions, and the magnitude of the charges on the ions. **(Ionic bonds)**

9. In NaF, the charges on the ions are 1+ and 1–. In MgF$_2$, the charges are 2+ and 1–. The higher the magnitude of the charges the greater the energy of the bond. A higher bond energy means that a greater force of attraction exists between the ions. Therefore, the force of attraction between ions is greater in MgF$_2$ than in NaF. **(Ionic bonds)**

10. Ionic compounds are hard but brittle crystalline solids with relatively high melting and boiling points. They generally have low electrical conductivities as solids, but are good electric conductors in the liquid (molten) state. Ionic compounds produce ions when dissolved in aqueous solutions. **(Ionic bonds)**

11. a. The bond order in a Cl$_2$ molecule is 1 because two electrons are shared.

    b. The bond order in a N$_2$ molecule is 3 because six electrons are shared.

    c. The bond order in a O$_2$ molecule is 2 because four electrons are shared. (Bond properties)

12. C≡C > C=C > C–C (bond energy) As the bond order decreases the bond energy decreases.

    C–C > C=C > C≡C (bond distance) As the bond order increases the bond distance decreases. **(Bond properties)**

13. a. Triple bonds are shorter than double bonds due to the greater electrostatic attractive forces. Hence, N≡N and C≡C are shorter than N=N and C=C. Because N and C are second period atoms, the atomic radius of N is smaller than that of C since sizes of atoms decrease across a period. Therefore, N≡N is shorter than C≡C and N=N is shorter than C=C. This means that the shortest bond distance is N≡N followed by C≡C. The next in the series is N=N and the longest is C=C.

$$N≡N < C≡C < N=N < C=C$$

    b. Bond energy is indirectly related to the bond distance. Shorter bonds have higher bond energies than longer ones. Thus, the order of bond energies is the opposite to that of bond distances. **(Bond properties)**

14. $\Sigma D_{reactants} = +432$ kJ + (+155 kJ) = 587 kJ

    $\Sigma D_{products} = 2 \times 565$ kJ/mol = 1130 kJ

    $\Delta H = \Sigma D_{reactants} - \Sigma D_{products} = 587$ kJ $- 1130$ kJ $= -543$ kJ

    The ΔH for the reaction is -543 kJ, an exothermic reaction. **(Bond properties)**

15. $\mu = q \times d = 1.60 \times 10^{-19}$ C $\times 1.27 \times 10^{-10}$ m $\times$ (1 D/ 3.34×10^{-30} C m) = 6.08 D

    % ionic character = (1.07 D/6.08 D) $\times$ 100 = 19.6% **(Bond properties)**

16. **(Lewis structures)**

    :S=C=S:

70 / College Chemistry

17. **(Lewis structures)**

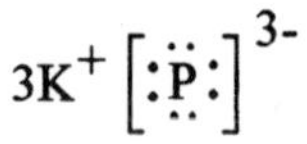

$$3K^+ \left[:\ddot{\underset{..}{P}}: \right]^{3-}$$

18. **(Lewis structures)**

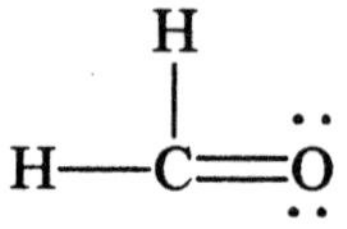

$$\text{H}-\underset{\underset{\text{H}}{|}}{\text{C}}=\ddot{\underset{..}{\text{O}}}$$

19. **(Lewis structures)**

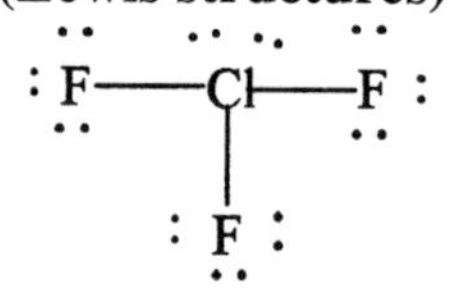

$$:\ddot{\underset{..}{F}}-\underset{\underset{:\ddot{\underset{..}{F}}:}{|}}{\overset{..}{\text{Cl}}}-\ddot{\underset{..}{F}}:$$

20. **(Lewis structures)**

$$:N\equiv N-\ddot{\underset{..}{O}}: \longleftrightarrow :\ddot{N}=N=\ddot{O} \longleftrightarrow :\ddot{\underset{..}{N}}-N\equiv O$$

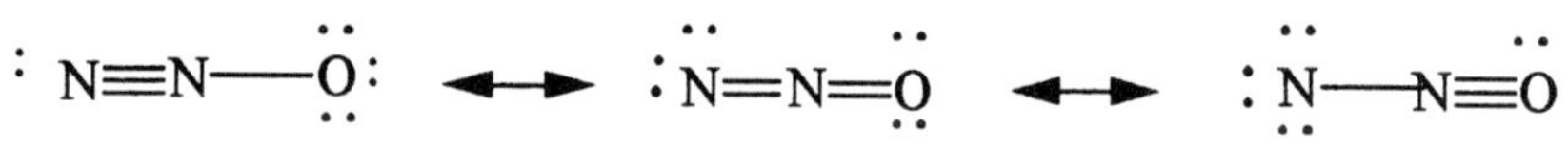

21. The formal charges on each atom in H–C≡N: is as follows.

    $FC_H = 1 - (0 + ½(2)) = 0$

    $FC_C = 4 - (0 + ½(8)) = 0$

    $FC_N = 5 - (2 + ½(6)) = 0$

    The formal charges on each atom in H–N≡C: is as follows.

    $FC_H = 1 - (0 + ½(2)) = 0$

    $FC_C = 4 - (2 + ½(6)) = -1$

    $FC_N = 5 - (0 + ½(8)) = +1$

    H–C≡N: is a better prediction for the molecular structure of hydrogen cyanide than the other because all of the atoms have a formal charge of zero. **(Formal charges)**

22. :N≡C–B̈r:, each of its atoms has a zero formal charge, more plausible

    :C≡N–B̈r:, the formal charges on C and N are –1 and +1, respectively, less plausible

    (Formal charges)

23. a. Molecules exhibit resonance whenever more than one correct Lewis structure can be written that differs only in the placement of electrons. b. Each Lewis structure that differs with respect to the placement of the electrons in a molecule is called a contributing or resonance structure. c. Each contributing structure is separated with a double headed arrow. (↔) **(Resonance)**

24. Calcium oxide is an ionic compound. The Ca^{2+} and O^{2-} ions are in a three-dimensional crystal lattice. No molecules are found in this structure. Ionic compounds have formula units. **(Ionic bonds)**

25. When ionic solids dissolve the ionic bonds that hold the ions are broken and the water molecules surround and bond to the ions. The lattice energy for $Ca(OH)_2$ is much higher than that of KOH because of the higher charge on the Ca^{2+} ion. Hence, at 20°C sufficient energy is available to break the weaker ionic

bonds in KOH, but not enough energy is available to break the stronger ionic bonds in Ca(OH)$_2$. **(Ionic bonds)**

26. a. The NO$_2^-$ ion exhibits resonance. The two principal resonance structures differ in the placement of the N–O double and single bonds (O=N–O and O–N=O). This means that the actual bond is the average of the two. b. The bond order is 1.5. **(Bond properties)**

27. P–F > C–F > N–F = Cl–F The polarity of a bond is determined from the difference in electronegativity. The P–F bond has the greatest difference and the N–F and Cl–F bonds have the smallest. **(Bond properties)**

28. The value of the dipole moment depends on the magnitude of the charge separation and the distance that separates the charges. Because F (4.0) is significantly more electronegative than Cl (3.0) the charge separation in H–F is much greater than that in H–Cl. Thus, H–F has a greater dipole moment then H–Cl. **(Bond properties)**

29. $\mu = q \times d = 1.60 \times 10^{-19}$ C $\times 1.13 \times 10^{-10}$ m $\times$ (1 D/ 3.34×10^{-30} C m) = 5.41 D

    % ionic character = 2.0% = $(x/5.41$ D$) \times 100$

    $x = 0.020 \times 5.41$ D = 0.11 D **(Bond properties)**

30. The C–C triple bond and Br–Br single bond are broken, and a C–C double bond and C–Br single bond are formed.

    $\Sigma D_{reactants}$ = 839 kJ + 193 kJ = 1032 kJ

    $\Sigma D_{products}$ = (2 × 276 kJ/mol) + 614 kJ = 1166 kJ

    $\Delta H = \Sigma D_{reactants} - \Sigma D_{products}$ = 1032 kJ – 1166 kJ = –134 kJ **(Bond properties)**

31. **(Lewis structures)**

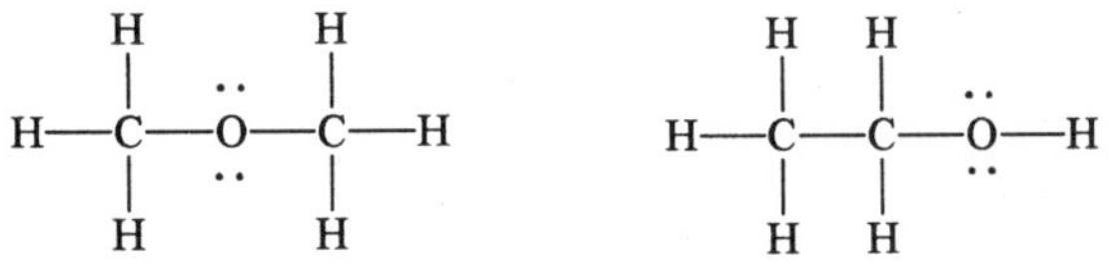

32. **(Lewis structures)**

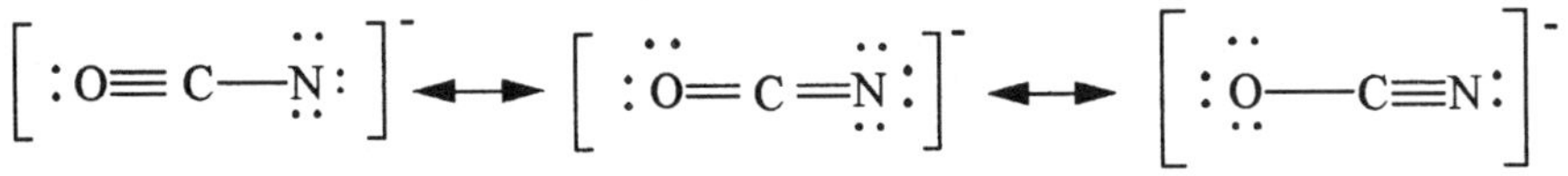

33. **(Lewis structures)**

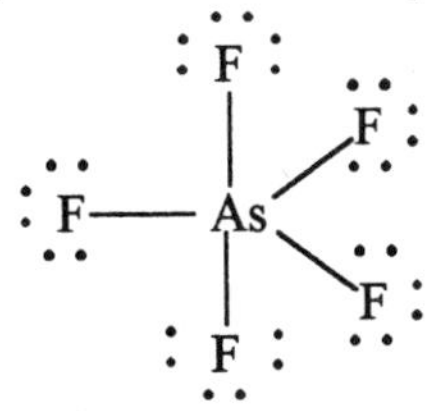

34. **(Lewis structures)**

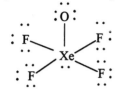

35. a. The size of the Cs⁺ cation in these cesium salts is fixed but the size of the halide ions vary. F⁻ is the smallest and I⁻ is the largest. The longest bond is in CsI and the shortest is in CsF. Hence, CsI has the lowest lattice energy and CsF has the highest. The longer the distance between ions with the same charges, the weaker the bond. **(Ionic bonds)**

36. d. ClI has the longest bond because the Cl atom is bonded to the largest halogen atom I. The overlap in this molecule is the 3*p* from Cl with the 5*p* of I. All of the other bonds are shorter because of the smaller size of the overlapping orbitals: 2*p* in F, 3*p* in Cl, and 4*p* in Br. **(Bond properties)**

37. a. O–F < C=O > N≡N The strongest bond, the one with the highest bond energy, is N≡N because of the strong triple bond. The weakest is O–F because it only has a single bond. The intermediate one is C=O. **(Bond properties)**

38. b. HF has the greatest difference in electronegativity; thus, it has the greatest separation of charge and dipole moment. For this group of hydrogen halides, the electronegativity difference is a more important factor than the larger size of the atoms. **(Bond properties)**

39. d. Percent ionic character is related to the difference in electronegativity, ΔEN. O–I has the greatest ΔEN; hence, it has the largest percent ionic character. **(Bond properties)**

40. a. 824 kJ

$K(s) + \frac{1}{2}F_2(g) \to KF(s)$	$\Delta H = 563$ kJ
$K(s) \to K(g)$	$\Delta H = 90$ kJ
$K(g) \to e^- + K^+(g)$	$\Delta H = 415$ kJ
$\frac{1}{2}F_2(g) \to F(g)$	$\Delta H = 0.5 \times 155$ kJ
$F(g) + e^- \to F^-(g)$	$\Delta H = -322$ kJ
$KF(s) \to K^+(g) + F^-(g)$	$\Delta H = 824$ kJ

    (Bond properties)

Grade Yourself

Circle the numbers of the questions you missed, then fill in the total incorrect for each topic. If you answered more than three questions incorrectly, you need to focus on that topic. (If a topic has less than three questions and you had at least one wrong, we suggest you study that topic also. Read your textbook, a review book, or ask your teacher for help.)

Subject: Chemical Bonds—Fundamental Concepts

Topic	Question Numbers	Number Incorrect
Chemical bonds	1, 2	
Ionic bonds	3, 4, 5, 6, 7, 8, 9, 10, 24, 25, 35	
Bond properties	11, 12, 13, 14, 15, 26, 27, 28, 29, 30, 36, 37, 38, 39, 40	
Lewis structures	16, 17, 18, 19, 20, 31, 32, 33, 34	
Formal charges	21, 22	
Resonance	23	

The Structure of Molecules and Molecular Orbitals

Brief Yourself

Molecular geometries are predicted by using the valence shell electron pair repulsion (VSEPR) method. This method is based on the fact that the electron pairs on the central atom repel each other so that they are as far as possible from each other. A molecule that has a central atom with two bonding and no lone pairs has a linear geometry because the bonding orbitals are at a maximum distance from each other when the bond angle is 180°. Three electron pairs on the central atom have trigonal planar geometry (120°). Four electron pairs have tetrahedral geometry (109.5°), five pairs have trigonal bipyramidal geometry (90°, 120°), and six pairs have octahedral geometry (90°). Lone pairs on the central atoms depress the bonding angles from their theoretical maximum values.

Valence bond theory explains the formation of bonds and molecular geometry in terms of hybrid orbitals on the central atom. A hybrid orbital results when two or more atomic orbitals are combined mathematically. If one s orbital and three p orbitals hybridize, four degenerate sp^3 hybrid orbitals result that have tetrahedral geometry. If one s and two p orbitals hybridize, three degenerate sp^2 hybrid orbitals result that have trigonal planar geometry. In addition, an unhybridized p orbital is located perpendicular to the plane of the sp^2 orbitals. When an s orbital hybridizes with a p orbital, two sp hybrid orbitals result that have a linear geometry. Additionally, two p orbitals remain unhybridized mutually perpendicular to each other and to the other bonds. Molecules that have central atoms with more than eight electrons have hybrid orbitals that combine s, p, and d orbitals. Central atoms with 10 electrons principally use sp^3d, and central atoms with 12 electrons mainly use sp^3d^2 hybrid orbitals.

The molecular orbital theory explains the formation of bonds in terms of the linear mathematical combinations of the wave functions for all of the atomic orbitals. When two atomic orbitals combine they produce two molecular orbitals—a lower energy bonding molecular orbital and a higher energy antibonding molecular orbital. The bonding orbital has a large electron density between the two nuclei, and the antibonding orbital has minimal electron between the two nuclei. The sum of the energies of the bonding and the antibonding molecular orbitals equals the sum of the energies of the atomic orbitals. The combination of two 1s orbitals produces the σ_{1s} and σ^*_{1s} and the combination of two 2s orbitals produces the two σ_{2s} and σ^*_{2s}. To obtain molecular orbital configurations, electrons from atomic orbitals are placed into molecular orbitals. Electrons in molecular orbitals follow the aufbau principle, Pauli's exclusion principle, and Hund's rule.

Test Yourself

1. For each of the following give the maximum angle between atoms around a central atom that has no lone pairs. a. two bonded atoms b. three bonded atoms c. four bonded atoms d. five bonded atoms e. six bonded atoms

2. What is the geometry of molecules with the following general formulas? The central atom is A. The atoms bonded to the central atom are designated by B, and E represents the lone pair electrons. a. AB_3 b. AB_3E c. AB_3E_2 d. AB_6 e. AB_2

3. Give an example of a molecule for each of the following general formulas and determine its molecular geometry. a. AB_4 b. AB_2E_2 c. AB_2E

4. a. What is the molecular geometry of silicon tetrachloride molecules? b. What is the bond angle in this molecule?

5. a. What is the molecular geometry of arsenic trifluoride molecules? b. What is the bond angle in this molecule?

6. a. What is the molecular geometry of bromine trifluoride molecules? b. What is the bond angle in this molecule?

7. a. What is the molecular geometry of carbon disulfide molecules? b. What is the bond angle in this molecule?

8. a. What is the molecular geometry of carbonate ions? b. What is the bond angle in this ion?

9. a. Explain why the H–N–H bond angle in NH_3 is less than 109.5°. b. Explain why the H–O–H bond angle in H_2O is smaller than the H–N–H bond angle in NH_3.

10. Compare the molecular geometries of ICl_2^- and ICl_4^-.

11. What are the molecular geometry and bond angles in C_3H_4?

12. a. What is the molecular geometry of IF_5? b. Is IF_5 a polar or nonpolar molecule?

13. What hybrid orbital is associated with a central atom that has four single bonds and no lone pairs?

14. What is the hybridization of the C atoms in ethylene, C_2H_4?

15. Three I atoms can form both a cation, I_3^+, and an anion, I_3^-. Predict the shapes of both ions.

16. How do σ and π bonds differ?

17. What changes in hybridization occur when Cl_2 reacts with C_2H_2 as follows?

$$C_2H_2 + Cl_2 \rightarrow C_2H_2Cl_2$$

$$C_2H_2Cl_2 + Cl_2 \rightarrow C_2H_2Cl_4$$

18. What is the hybridization of the orbitals of Se in SeF_6?

19. What is the hybridization of the orbitals around the central C atoms in $N\equiv C–C\equiv N$?

20. Boron trifluoride, BF_3, reacts with the fluoride ion, F^-, and produces the tetrafluoroborate ion, BF_4^-. Describe the change in hybridization and geometry that occurs in this reaction.

21. Allene, C_3H_4, is an unstable molecule composed of a chain of three C atoms in which the four H atoms bond to the C atoms at the end of the chain. What is the hybridization of each C atom in allene? b. What is the geometry of the allene molecule?

22. Sulfuryl fluoride, SO_2F_2, is an unreactive gas at 25°C. Analysis of the sulfuryl fluoride molecule reveals that the central S atom does not have any lone pair electrons and no multiple bonds are in the molecule. a. What is the geometry of a SO_2F_2 molecule? b. What hybrid orbitals does the central S atom use?

23. Describe the structure of the $AsCl_4^+$ ion.

24. Explain the trend in bond angles for the period 3 hydrides from groups 14 (IVA) to 16 (VIA).

Molecule	Bond Angle
SiH_4	109.5°
PH_3	93.8°
H_2S	92.2°

25. a. How does the σ_{1s} differ in energy from the σ^*_{1s}? b. How do these two orbitals differ with respect to their electron distribution?

26. Use the molecular orbital theory to explain why diatomic He, He_2, has never been observed.

27. a. Write the molecular orbital configuration for Li_2. b. What is the bond order of Li_2?

28. a. Use a molecular orbital diagram to determine the bond order and magnetic properties of a molecular oxygen cation, O_2^+. b. Compare the bond order of O_2^+ to the bond order of O_2.

29. Use the molecular orbital theory to rank N_2, N_2^+, and N_2^- in increasing order of bond energy.

30. Acetylide ions have the formula of C_2^{2-}. a. Use a MO diagram to predict the bond order of the acetylide ion. b. What molecule is the acetylide ion isoelectronic to?

31. Write the complete molecular orbital configuration for F_2.

32. Use the molecular orbital theory to explain why the first ionization energy of O_2 is lower than that of an O atom, but the first ionization energy of H_2 is higher than that of a H atom.

33. a. What is the molecular orbital configuration of NO? b. What is the bond order in NO?

34. a. Write the valence molecular-orbital configuration for Cl_2. b. What is the Cl–Cl bond order? c. Describe the magnetic property of Cl_2.

35. Which of the following molecules is linear?
    a. H_2S
    b. HCN
    c. OCl_2
    d. NF_3
    e. none of these

36. Which of the following molecules is square planar?
    a. SiF_4
    b. SF_4
    c. CF_4
    d. XeF_4
    e. none of these

37. Which of the following molecules is polar?
    a. CO_2
    b. NF_3
    c. SO_3
    d. XeF_2
    e. none of these

38. Which of the following has a central atom with sp^2 hybrid orbitals?

   a. BF_3
   b. IF_3
   c. OF_2
   d. CBr_4
   e. none of these

39. Which of the following has a central atom with a sp^3d hybrid orbital?

   a. NH_3
   b. IF_5
   c. SF_6
   d. XeF_4
   e. none of these

40. What is the hybridization of the P atom in PCl_4^+?

   a. sp
   b. sp^2
   c. sp^3
   d. sp^3d
   e. sp^3d^2

Check Yourself

1. a. 180°, b. 120°, c. 109.5°, d. 90° and 180°, e. 90° **(VSEPR method)**

2. a. trigonal planar, b. trigonal pyramidal, c. T-shaped, d. octahedral, e. linear **(VSEPR method)**

3. a. CH_4, tetrahedral, b. H_2O, angular, c. O_3, angular **(VSEPR method)**

4. a. tetrahedral, b. 109.5° The central Si atom has four single bonds to the four Cl atoms. **(VSEPR method)**

5. a. trigonal pyramidal, b. < 109.5° The central As atom has single bonds to the three F atoms and has one lone pair. **(VSEPR method)**

6. a. T-shaped, b. 90° The central Br atom forms single bonds with the three F atoms and has two lone pairs. **(VSEPR method)**

7. a. linear, b. 180° The central C atom forms double bonds to the two S atoms and has no lone pairs. **(VSEPR method)**

8. a. trigonal planar, b. 120° The central C atom in the carbonate ion, CO_3^{2-}, bonds to two O atoms with single bonds and one with a double bond (resonance hybrid). Thus, it has three bonds and no lone pairs. **(VSEPR method)**

9. a. The lone pair on the N atom repels the three N–H bonds below the theoretical maximum of 109.5° because lone pair-bonding pair repulsions are stronger than bonding pair-bonding pair repulsions. b. The O atom in water has two lone pairs that can depress the angle to a greater degree than the one lone pair on the N in ammonia. **(VSEPR method)**

10. The central I atom in ICl_2^- has ten valence electrons. Because it has two bonds and three lone pairs, the geometry of ICl_2^- is linear. The central I atom in ICl_4^- has twelve valence electrons. Because it has four bonds and two lone pairs, the geometry of ICl_4^- is square planar. **(Molecular structure)**

11. Both isomers of C_3H_4, $H_2C=C=CH_2$ and $H_3C-C\equiv C-H$, are linear because of having two bonds to the central C atom and no lone pairs. The bond angle associated with linear geometry is 180°. **(Molecular structure)**

12. a. square pyramidal, b. polar IF_5 because the central I atom has five bonds and one lone pair. It is polar because each bond is polar and the IF_5 molecule is not symmetrical. **(Molecular structure)**

13. Central atoms with four single bonds and no lone pairs have sp^3 hybrid orbitals. **(Hybridization)**

14. The central C atoms in C_2H_4 are sp^2 hybridized because of the double bond between them. **(Hybridization)**

15. The central I atom in I_3^+ has eight valence electrons, two bonds and two lone pairs; therefore I_3^+ is angular with bond angles less than 109.5°. The central I atom in I_3^- has ten valence electrons, two bonds and three lone pairs; therefore I_3^- is linear with a 180° bond angle. **(VSEPR method)**

16. A sigma bond, σ, forms when the region of maximum electron density is between the two nuclei of the bonding atoms. This is sometimes called end-on overlap of orbitals. A pi bond, π, forms when region of maximum electron density is above and below the two nuclei of the bonding atoms. This is sometimes called sideways overlap of orbitals. **(Multiple bonds)**

17. As the result of having a C–C triple bond, the C atoms in C_2H_2 are sp hybridized. When the first Cl_2 reacts, it breaks one of the π bonds and produces a C–C double bond that has sp^2 hybridized C orbitals. After the second Cl_2 reacts and breaks the other π bond, a C–C single bond results that has sp^3 hybridized C orbitals. **(Hybrid orbitals)**

18. The molecular geometry of SeF_6 is octahedral. Molecules with octahedral geometry have central atoms with sp^3d^2 hybrid orbitals. (Hybrid orbitals)

19. The Lewis structure shows that the C–N bonds are triple bonds and the C–C bond is a single bond. Thus, the two C atoms have two σ bonds and two π bonds. Two π bonds require two unhybridized p orbitals. Therefore, the C atoms have two sp hybrid orbitals to form the two σ bonds and the remaining two unhybridized p orbitals overlap with two p orbitals on the N atoms. **(Hybrid orbitals)**

20. The B in BF_3 is sp^2 hybridized and it changes to sp^3 when BF_4^- forms. **(Hybrid orbitals)**

21. The Lewis structure of allene is $H_2C=C=CH_2$. a. The C atoms on the end form a double bond with the central C atom; thus, they are sp^2 hybridized. The central C atom forms two double bonds; hence, it forms sp hybrid orbitals. b. The molecule is linear which is consistent with the sp hybrid orbitals on the central C atom. **(Hybrid orbitals)**

22. a. The SO_2F_2 molecule is tetrahedral because the S atom has four bonds and no lone pairs. b. The hybrid orbitals used by S is sp^3. **(Hybrid orbitals)**

23. The As atom in $AsCl_4^+$ has eight electrons in four bonds to the Cl atoms; thus, it is sp^3 hybridized with tetrahedral geometry. **(Molecular structure)**

24. The Si atom in SiH_4 has four bonds with no lone pairs; thus, the bond angle is 109.5°. The P atom in PH_3 has three bonds and one lone pair, which depresses the bond angle to 93.8°. The S atom in H_2S has two bonds and two lone pairs, which depresses the bond angle to 92.2°. Two lone pairs can depress the angle to a greater degree than just one lone pair. **(Molecular structure)**

25. a. The σ_{1s}, a bonding orbital, is lower in energy than the σ^*_{1s}, an antibonding orbital. b. Bonding orbitals have a large electron density between the two nuclei and antibonding orbitals have little electron density between the nuclei. This means the nuclei repel and thus are "anti" bonding. **(Molecular orbitals)**

26. The molecular orbital configuration for He$_2$ is $(\sigma_{1s})^2 (\sigma^*_{1s})^2$. Whenever the number of electrons in bonding orbitals equals those in antibonding orbitals, the bond order is zero, which means that no bond exists. In other words, the two He atoms are more stable as isolated atoms as a result of having electrons in the lower energy 1s. **(Molecular orbitals)**

27. a. $(\sigma_{1s})^2 (\sigma^*_{1s})^2 (\sigma_{2s})^2$

    b. Bond order Li$_2$ = (bonding e$^-$ – antibonding e$^-$)/2 = (4 e$^-$ – 2e$^-$)/2 = 1 **(Molecular orbitals)**

28. a. The molecular orbital configuration for O$_2^+$ is $(\sigma_{1s})^2 (\sigma^*_{1s})^2 (\sigma_{2s})^2 (\sigma^*_{2s})^2 (\sigma_{2p})^2 (\pi_{2p})^4 (\pi^*_{2p})^1$.

    b. Bond order O$_2^+$ = (bonding e$^-$ – antibonding e$^-$)/2 = (10 e$^-$ – 5 e$^-$)/2 = 2.5 b. Removing an antibonding electron from O$_2$ makes a stronger bond than that in O$_2$ which has a bond order of 2. **(Molecular orbitals)**

29. The molecular orbital configuration for N$_2$ is $(\sigma_{1s})^2 (\sigma^*_{1s})^2 (\sigma_{2s})^2 (\sigma^*_{2s})^2 (\pi_{2p})^4 (\sigma_{2p})^2$, giving a bond order of 3. The molecular orbital configuration for N$_2^+$ is $(\sigma_{1s})^2 (\sigma^*_{1s})^2 (\sigma_{2s})^2 (\sigma^*_{2s})^2 (\pi_{2p})^4 (\sigma_{2p})^1$, giving a bond order of 2.5. The molecular orbital configuration for N$_2^-$ is $(\sigma_{1s})^2 (\sigma^*_{1s})^2 (\sigma_{2s})^2 (\sigma^*_{2s})^2 (\pi_{2p})^4 (\sigma_{2p})^2 (\pi^*_{2p})^1$, giving a bond order of 2.5. Thus, N$_2^+$ and N$_2^-$ have a lower bond energy than N$_2$ because bond energy is directly related to bond order. **(Molecular orbitals)**

30. a. The molecular orbital configuration of C$_2^{2-}$ is $(\sigma_{1s})^2 (\sigma^*_{1s})^2 (\sigma_{2s})^2 (\sigma^*_{2s})^2 (\pi_{2p})^4 (\sigma_{2p})^2$, giving a bond order of 3. b. This is the same molecular orbital configuration as N$_2$. **(Molecular orbitals)**

31. $(\sigma_{1s})^2 (\sigma^*_{1s})^2 (\sigma_{2s})^2 (\sigma^*_{2s})^2 (\sigma_{2p})^2 (\pi_{2p})^4 (\pi^*_{2p})^4$ **(Molecular orbitals)**

32. The first ionization energy is the minimum energy needed to remove an electron. The outermost electron in O$_2$ is in a higher energy orbital, π^*_{2p}, than the O atom, 2p; thus, the first ionization energy of O$_2$ is less than that of O. In contrast, the outermost electron in H$_2$, σ_{1s}, is in a lower energy orbital than a H atom, 1s; hence, more energy is needed to remove the electron in the σ_{1s} orbital and the 1s. **(Molecular orbitals)**

33. a. $(\sigma_{1s})^2 (\sigma^*_{1s})^2 (\sigma_{2s})^2 (\sigma^*_{2s})^2 (\sigma_{2p})^2 (\pi_{2p})^4 (\pi^*_{2p})^1$

    b. Bond order$_{NO}$ = (bonding e$^-$ – antibonding e$^-$)/2 = (10 e$^-$ – 5e$^-$)/2 = 2.5 **(Molecular orbitals)**

34. a. $(\sigma_{3s})^2 (\sigma^*_{3s})^2 (\sigma_{3p})^2 (\pi_{3p})^4 (\pi^*_{3p})^4$

    b. Bond order Cl$_2$ = (bonding e$^-$ – antibonding e$^-$)/2 = (8 e$^-$ – 6e$^-$)/2 = 1

    c. Because all of the electrons in the molecular orbitals of Cl$_2$ are paired, Cl$_2$ is a diamagnetic molecule. **(Molecular orbitals)**

35. b. The HCN molecule has a H–C single bond and a C–N triple bond; thus, the molecule is linear. **(VSEPR method)**

36. d. The Xe in XeF$_4$ has four bonds and two lone pairs; hence, it is square planar. **(VSEPR method)**

37. b. NF$_3$ is polar because each bond is polar and the molecule is not symmetrical. **(Molecular structure)**

38. a. The B atom in BF$_3$ is sp^2 hybridized because it has three B–F single bonds and no lone pairs. **(Hybrid orbitals)**

39. e. None of them have a combination of five bonds and/or lone pairs. **(Hybrid orbitals)**

40. c. The P atom in PCl$_4^+$ is sp^3 hybridized because it has eight valence electrons and no lone pairs. **(Hybrid orbitals)**

Grade Yourself

Circle the numbers of the questions you missed, then fill in the total incorrect for each topic. If you answered more than three questions incorrectly, you need to focus on that topic. (If a topic has less than three questions and you had at least one wrong, we suggest you study that topic also. Read your textbook, a review book, or ask your teacher for help.)

Subject: The Structure of Molecules and Molecular Orbitals

Topic	Question Numbers	Number Incorrect
VSEPR method	1, 2, 3, 4, 5, 6, 7, 8, 9, 15, 35, 36	
Molecular structure	10, 11, 12, 23, 24, 37	
Hybridization	13, 14	
Multiple bonds	16	
Hybrid orbitals	17, 18, 19, 20, 21, 22, 38, 39, 40	
Molecular orbitals	25, 26, 27, 28, 29, 30, 31, 32, 33, 34	

Liquids, Solids, and Changes of State

 ## Brief Yourself

When the pressure of a gas increases and its temperature decreases, the intermolecular forces overcome the kinetic energy of the gas particles and the gas condenses to a liquid. Dipole-dipole interactions, hydrogen bonds, and London dispersion forces are the three principal intermolecular forces in pure liquids. Dipole-dipole interactions are short-range forces between the molecules that have a permanent dipole. Hydrogen bonds are dipole-dipole forces among molecules that have a H atom bonded to one of the three most electronegative atoms—F, O, or N. These electronegative atoms produce a large charge separation because removal of electron density from the H atom exposes the nucleus, a proton, which has a high charge density. Hence, H bonds are significantly stronger than dipole-dipole interactions. London dispersion forces are in all liquids but they are most important in nonpolar liquids. When the distance between two nonpolar molecules is small, the electron cloud from one molecule induces an instantaneous dipole in the other, resulting in very weak attractive forces between the two molecules.

Evaporation of liquids occurs when higher-energy surface molecules break free and enter the vapor phase. In a closed system, a liquid reaches a point when the rate of evaporation equals the rate of condensation. At this point a dynamic equilibrium exists. The pressure exerted by a vapor in equilibrium with a liquid is termed equilibrium vapor pressure. As the temperature increases, the vapor pressure of a liquid increases because more molecules have sufficient energy to overcome the forces that bind them to the liquid. The temperature at which the vapor pressure equals the external pressure is called the boiling point. The molar enthalpy of vaporization, ΔH_{vap}, must be added to evaporate one mole of a liquid at a constant temperature. Liquids with stronger intermolecular forces have higher boiling points and molar enthalpies of vaporization than those with weaker intermolecular forces.

Liquids behave as if they have a membrane stretched across their surface. This behavior results from the unbalanced forces on the surface molecules. A measure of this effect is called surface tension, which is the energy required to increase the surface area of a liquid by a unit amount. A measure of the resistance of a liquid to flow is its viscosity.

Solids have constant volumes and shapes, do not exhibit fluid properties, and are incompressible. Crystalline solids have a regular pattern of particles. A rather small number of solids, amorphous solids, have a more random arrangement of particles that more closely resembles the structures of liquids. Solids are classified according to the particles in their crystal lattices and the type of forces among the particles. Ions, bonded by ionic bonds, are at the crystal lattice points of ionic solids. Covalent solids, also called network or macromolecular solids, have covalently bonded atoms at their crystal lattice points. Molecular solids have molecules at their crystal lattice points bound by dipole-dipole interactions, hydrogen bonds, and London forces. Metallic solids have nuclei of metal atoms surrounded by delocalized electrons. The electrostatic forces of attraction of these nuclei and electrons are called metallic bonds.

Crystalline solids have an orderly array of particles called a crystal lattice. The smallest repeating pattern of particles in the structure is the unit cell. Cubic, tetragonal, orthorhombic, monoclinic, triclinic, rhombohedral, and hexagonal are the seven unit cells found in crystalline solids. In addition, some unit cells have particles inside and along the edges; e.g., face-centered cubic and body-centered cubic. Particles in crystalline solids tend to occupy as little volume as possible. The greater the attractive forces, the closer particles pack in the crystal structure. Hexagonal-closest packing and cubic-closest packing are the two closest-packed structures most commonly found in metals and other solids. X-ray diffraction analysis is used to determine the structure of solids because x-rays with wavelengths about the same length as the spacing between layers of atoms are reflected at fixed angles.

When a solid is heated its temperature increases until the melting point is reached. At the melting point, some of the bonds in the solid break and the liquid state results. During melting, the temperature remains constant. After all of the solid changes to a liquid, addition of heat causes its temperature to rise until its boiling point is reached. At the boiling point, all of the intermolecular forces in the liquid break and the molecules enter the vapor phase. Heating and cooling curves are plotted to show the behavior of substances as heat is added or removed. A phase diagram shows the physical states of substance at a given temperature and pressure, and the conditions under which it changes state.

Test Yourself

1. Contrast dipole-dipole interactions and London dispersion forces in liquids.

2. What are the principal intermolecular forces in $Kr(l)$?

3. What are the principal intermolecular forces in $H_2S(l)$?

4. What are the principal intermolecular forces in $CH_3CH_2OH(l)$?

5. Consider the boiling points of $HCl(l)$, $HBr(l)$, and $HI(l)$:

Compound	Boiling Point, °C (1 atm)
HCl	−85.0
HBr	−66.7
HI	−35.4

   Explain the increasing boiling points of these liquids in terms of their intermolecular forces.

6. Consider the following liquids: $CH_3OH(l)$, $CH_3CH_2OH(l)$, $CH_4(l)$, $CH_3CH_3(l)$. Arrange these liquids from lowest to highest boiling point.

7. a. What is molar enthalpy of vaporization? b. How is molar enthalpy of vaporization linked to the strength of the intermolecular forces in liquids?

8. a. Compare the vapor pressures of water and diethylether, $(C_2H_6)_2O$, at room temperature. b. Which is a more volatile liquid?

9. What is the critical temperature and critical pressure of a liquid?

10. At 233.0 K, liquid ammonia, $NH_3(l)$, has a vapor pressure of 0.7083 atm and at 221.0 K its vapor pressure is 0.3578 atm. Calculate the molar enthalpy of vaporization, ΔH_{vap}, of liquid ammonia in kJ/mol.

11. Explain why the boiling point of water is lower at higher altitudes than it is at sea level.

12. Predict the class of solid to which each of the following belongs: $KNO_3(s)$, $I_2(s)$, $SiC(s)$, and $Fe(s)$.

13. What are the general properties of molecular solids?

14. a. What are unit cells in crystalline solids?
    b. What is their importance?

15. Describe the properties of the simple cubic unit cell.

16. Describe the main difference in structure of cubic, tetragonal, and orthorhombic unit cells.

17. Copper atoms have a face-centered cubic structure. The radius of a Cu atom is 128 pm. Calculate the volume of the unit cell in cm^3.

18. What is the difference between cubic closest packed and hexagonal closest packed structures?

19. An unknown crystal is analyzed using x-rays with a wavelength 165 pm. The first-order ($n = 1$) reflection angle is 11.3°. What distance separates the layers of atoms in the unknown crystal?

20. a. What is a crystal defect? b. Give an example.

21. Explain why the magnitude of the molar enthalpy of fusion of a substance is usually much smaller than its molar enthalpy of vaporization.

22. Explain why diethylether, $(CH_3CH_2)_2O$, feels cold when it is poured on your skin.

23. Benzene, C_6H_6, has a molar enthalpy of vaporization of 30.8 kJ/mol at its boiling point, 80.1°C. Calculate the quantity of heat that must be added to 156 g $C_6H_6(l)$ at 80.1°C to benzene vapor at the same temperature.

24. The vapor pressure of $N_2(l)$ is 109.7 torr at −209°C and the molar enthalpy of vaporization of N_2 is 5.58 kJ/mol. Calculate the normal boiling point of N_2.

25. The critical temperature and pressure of Ar is −122°C and 48 atm, respectively. What do these data tell about the properties of Ar?

26. In what general class of solids does each of the following belong: a. silicon dioxide, $SiO_2(s)$ b. calcium, $Ca(s)$ c. iron(III) oxide, $Fe_2O_3(s)$ d. tetraphosphorus decoxide, $P_4O_{10}(s)$?

27. How many net spherical atoms are within a face-centered cubic unit cell? Explain.

28. If a 100-g sample of liquid water is initially at 0.0°C, calculate the amount of heat in kJ needed to produce 100 g $H_2O(g)$ at 100°C. The specific heat of water is 4.184 J/(g°C) and the molar enthalpy of vaporization of water is 40.8 kJ/mol.

29. The unit cell in silver is face-centered cubic and the edge length is 409 pm. Calculate the density of silver in g/cm^3.

(For Problems 30 to 34 use the following phase diagram substance Q.)

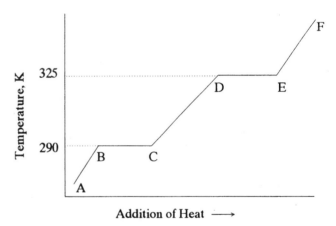

30. At what temperature does substance Q melt?

    a. 325 K

    b. 290 K

    c. 325 K

    d. 290-325 K

    e. cannot determine from the graph

31. Which segment of the curve shows when substance Q is in the liquid state?

    a. AB

    b. BC

    c. CD

    d. DE

    e. EF

32. Which segment of the curve shows the increase in temperature when solid Q is heated?

    a. AB

    b. BC

    c. CD

    d. DE

    e. EF

33. How much heat is required to go from point D to E on the heating curve for substance Q?

    a. specific heat of $Q(l)$

    b. specific heat of $Q(g)$

    c. enthalpy of fusion

    d. enthalpy of vaporization

    e. none of these

34. Which of the following best explains why heat is added from B to C but the temperature does not increase?

    a. The added heat breaks the bonds in solid Q causing an increase in the potential energy but does not increase the average kinetic energy of the particles.

    b. The added heat increases the average kinetic energy of the particles in solid Q.

    c. The added heat breaks the bonds in liquid Q causing increases in the potential energy and average kinetic energy of the particles.

    d. The added heat breaks the bonds in solid Q and increases the average kinetic energy of the particles.

    e. none of these

35. Which of the following is a hydrogen-bonded liquid?

    a. $HCl(l)$

    b. $NH_3(l)$

    c. $CH_4(l)$

    d. $SiH_4(l)$

    e. none of these

36. Which of the following has the highest boiling point?

    a. $PCl_3(l)$

    b. $PF_3(l)$

    c. $PBr_3(l)$

    d. $PH_3(l)$

    e. none of these

37. Which of the following has the lowest molar enthalpy of vaporization?

    a. $C_5H_{12}(l)$

    b. $C_6H_{14}(l)$

    c. $C_7H_{16}(l)$

    d. $C_8H_{18}(l)$

    e. none of these

38. If the vapor pressure of ethyl acetate is 0.395 atm at 51°C and 1.00 atm at 77°C, what is its molar enthalpy of vaporization.

    a. 1.17 kJ

    b. 3.37×10^4 kJ

    c. 14.6 kJ

    d. 33.7 kJ/mol

    e. none of these

39. Which of the following solids has the highest melting point?

    a. $Mg(s)$

    b. $MgCl_2(s)$

    c. $SiO_2(s)$

    d. $CO_2(s)$

    e. $H_2O(s)$

(For Problems 40 to 45 use the following phase diagram substance X.)

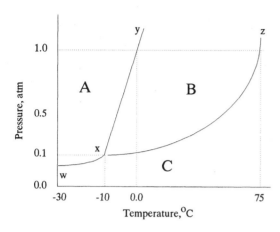

40. In what physical state is substance X in region B?
   a. solid
   b. liquid
   c. gas
   d. plasma
   e. solution

41. In what physical state is substance X at 0.5 atm and 80°C?
   a. solid
   b. liquid
   c. gas
   d. plasma
   e. solution

42. What is the meaning of line xy?
   a. It shows the temperature that substance X melts.
   b. It shows the temperature that substance X freezes.
   c. It shows the temperature that substance X boils.
   d. It shows the temperature that substance X sublimes.
   e. Both a and b are correct

43. At what temperature does substance X boil at 1.0 atm?
   a. 100°C
   b. 0.0°C
   c. 75°C
   d. –10°C
   e. none of these

44. Which of the following describes the state changes that occur in substance X when the pressure is increased from 0.01 atm to 1.5 atm at 0.0°C
   a. At 0.1 atm the vapor condenses to a liquid and at 1.0 atm it freezes.
   b. At 0.1 atm the liquid boils and at 1.0 atm it condenses.
   c. At 0.1 atm the vapor condenses to a solid and at 1.0 atm it melts.
   d. At 0.1 atm the solid melts and at 1.0 atm it boils.
   e. none of these

45. At what set of conditions does substance X exist as a solid, liquid, and gas?
   a. 0.1 atm, 100°C
   b. 0.1 atm, 10°C
   c. 0.1 atm, 0.0°C
   d. 0.1 atm, –30°C
   e. none of these

Check Yourself

1. The molecules in liquids that attract by dipole-dipole forces have charge separations; e.g., HBr(l). The molecules in liquids that attract by London forces have no charge separation; e.g., $CCl_4(l)$. **(Intermolecular forces)**

2. Krypton, Kr, is a member of the group 18 elements, the noble gases. Kr atoms are neutral and do not have a charge separation. Hence, liquid Kr has London forces among the atoms. **(Intermolecular forces)**

3. Because S is in group 16 (VIA), it has six valence electrons. Therefore, it forms two single covalent bonds with the H atoms. Additionally, the S atom has two lone pairs. Atoms with two bonds and two lone pairs have angular geometry. The electronegativity of S (2.5) is greater than that of H (2.2); thus, the molecule is polar. Polar molecules form dipole-dipole forces. **(Intermolecular forces)**

4. Ethanol, $CH_3CH_2OH(l)$, has a H atom bonded to an O atom; therefore, it forms H bonds. The remaining part of the molecule, the ethyl group (CH_3CH_2-), is essentially nonpolar. **(Intermolecular forces)**

5. Going from HCl to HI, the number of polarizable electrons (more loosely-held electrons) increases which increases the strength of the London forces among molecules. With stronger London forces, higher temperatures are needed to vaporize the liquid, even though the strength of the dipole-dipole interactions decreases within this group. Additionally, as the molecular mass of a molecule increases, the faster it has to move to escape from the liquid. **(Intermolecular forces)**

6. $CH_3OH(l)$ and $CH_3CH_2OH(l)$ are hydrogen-bonded liquids. $CH_4(l)$ and $CH_3CH_3(l)$ are nonpolar molecules and have London forces. Hydrogen bonds are stronger than London forces and the larger a molecule the stronger the London forces. Thus, the correct order of boiling points is as follows.

$$CH_4 < CH_3CH_3 < CH_3OH < CH_3CH_2OH$$ **(Intermolecular forces)**

7. a. The energy required to evaporate one mole of a liquid at a constant temperature is the molar enthalpy of vaporization, ΔH_{vap}. ΔH_{vap} is the difference between the enthalpy of the vapor, H_{vapor}, and the enthalpy of the liquid, H_{liquid}. b. The magnitude of the molar enthalpy of vaporization is a good measure of the strength of the intermolecular forces in liquids. Liquids with higher values for ΔH_{vap} have stronger intermolecular forces than those with lower values. **(Liquid properties)**

8. a. Water is a hydrogen-bonded liquid which means it has a significantly lower vapor pressure than diethylether, $(C_2H_6)_2O$, which is essentially a nonpolar molecule with London forces. At 25°C, the vapor pressures of water and diethylether are 24 and 545 torr, respectively. b. Liquids such as diethylether with high vapor pressures are volatile liquids-those that evaporate readily. **(Liquid properties)**

9. The critical temperature is the temperature above which a gas cannot be condensed to a liquid by an increase in pressure. The critical pressure is the pressure required to condense a vapor at the critical temperature. **(Liquid properties)**

10. 8.314 J/(mol K) × (1 kJ/1000 J) = 0.008314 kJ/(mol K)

    $\ln P_2/P_1 = (\Delta H_{vap}/R)((t_2 - t_1)/(t_1 \cdot t_2))$

    $\ln(0.7083 \text{ atm}/0.3578 \text{ atm}) = (\Delta H_{vap}/0.008314 \text{ kJ/mol K}) \times ((233.0 \text{ K} - 221.0 \text{ K})/(221.0 \text{ K} \times 233.0 \text{ K}))$

    $0.6831 = (\Delta H_{vap}/0.008314 \text{ kJ/mol K}) \times (2.33 \times 10^{-4} \text{ K}^{-1})$

    $\Delta H_{vap} = 24.4$ kJ/mol **(Liquid properties)**

11. As the altitude increases, atmospheric pressure decreases. A liquid boils when the vapor pressure of the liquid equals the external pressure; thus, as the atmospheric pressure decreases, the boiling point decreases. **(Liquid properties)**

12. $KNO_3(s)$ is an ionic solid. $I_2(s)$ is a molecular solid. $SiC(s)$ is a covalent or network solid, and $Fe(s)$ is a metallic solid. **(Solid properties)**

13. Because the intermolecular forces in molecular solids (London forces, dipole-dipole forces, and hydrogen bonds) are weak compared to ionic and covalent bonds, molecular solids tend to be soft and have low to moderate melting points. Some molecular solids sublime (e.g., $CO_2(s)$ and $I_2(s)$) under normal atmosphere pressure. **(Solid properties)**

14. The repeating units that compose solids are called unit cells. A unit cell is the smallest repeating unit that determines the overall shape of a crystal. (**Structure of solids**)

15. A simple cubic pattern has equal edge lengths ($a = b = c$), and the angles between the edges all equal 90° ($\alpha = \beta = \gamma = 90°$), in which a, b, and c are lengths of the three edges and α, β, and γ are the angles between the edges. (**Structure of solids**)

16. Cubic, tetragonal, and orthorhombic unit cells all have 90° angles but differ in their edge lengths. In cubic structures the edge lengths are equal. In tetragonal structures two edge lengths are equal, and in orthorhombic none of the edge lengths are equal. (**Structure of solids**)

17. The Cu atom in the center of the face is tangent to the corner Cu atoms. Thus, the diagonal distance across the face equals four times the atomic radius (4×128 pm $= 512$ pm). The diagonal distance across the unit cell, 512 pm, is the hypotenuse of an isosceles right triangle in which the other two sides equal a, the edge length. Use the Pythagorean theorem, $a^2 + b^2 = c^2$ to calculate the edge length. In the equation, a^2 equals b^2, thus $a^2 + a^2 = c^2$ and c equals 512 pm, therefore

    $2a^2 = (512 \text{ pm})^2$

    $a = 362$ pm

    $V_{cube} = a^3 = (362 \text{ pm})^3 = 4.74 \times 10^7 \text{ pm}^3$

    $V_{cube} = 4.74 \times 10^7 \text{ pm}^3 \times (1 \text{ m}^3/1 \times 10^{36} \text{ pm}^3) \times (1 \times 10^6 \text{ cm}^3/1 \text{ m}^3) = 4.74 \times 10^{-23} \text{ cm}^3$

    (**Structure of solids**)

18. The arrangement of the first layer of particles is called A and the arrangement of the second layer in the gaps between the first layer is termed B. If the third layer of particles lie directly over the first layer, then the layers are represented as $ABABAB$. This packing pattern is hexagonal closest packing. When the third layer does not lie directly over either of the first two layers, it is known as the C layer. This gives the pattern $ABCABCABC$ in which every fourth layer lies directly over top of each other. This packing pattern is cubic closest packing. (**Structure of solids**)

19. $n\lambda = 2d \sin \theta$

    $d = n\lambda/(2 \sin \theta) = 1 \times 165 \text{ pm}/(2 \times \sin(11.3°)) = 421$ pm (**Structure of solids**)

20. a. Perfect crystals do not exist. Most real crystals have imperfections which are called crystal defects. This means that foreign atoms and ions occupy points in the crystal lattice. b. A ruby is a crystal with impurities in some of its lattice positions. The principal structure that makes up rubies is colorless aluminum oxide, Al_2O_3. However, the substitution of chromium(III) ions, Cr^{3+}, in the crystal for some Al^{3+} ions gives rise to the red color of rubies. (**Structure of solids**)

21. Because all of the intermolecular forces in a liquid must be broken for it to become a vapor, the magnitude of the molar enthalpy of vaporization is generally much larger than the molar enthalpy of fusion in which only some of the attractive forces are overcome to enter the liquid phase. (**Changes of state**)

22. Diethylether has weak dipole-dipole and London forces; therefore, it evaporates readily when on the surface of skin (37°C). Body heat increases the temperature of diethylether to its boiling point (34.6°C) and then provides the enthalpy of vaporization. The loss of heat causes the temperature of the skin to decrease. (**Changes of state**)

23. 156 g $C_6H_6 \times$ (1 mol C_6H_6/78.0 g) $\times$ (30.8 kJ/mol) = 61.6 kJ (**Changes of state**)

24. $\ln(P_2/P_1) = \Delta H_{vap}/R\,[(T_2 - T_1)/(T_1 T_2)]$

At the normal boiling point T_{bp} the vapor pressure is 760 torr.

$\ln(760\text{ torr}/109.7\text{ torr}) = (5.58\text{ kJ}/0.008314\text{ kJ/(mol K)}) \times ((T_{bp} - 64\text{K})/(64\text{ K} \times T_{bp}))$

$T_{bp} = 78\text{ K} = -195°\text{C}$ **(Liquid properties)**

25. At $-122°\text{C}$, the critical temperature, 48 atm is required to liquify Ar(g). Above $-122°\text{C}$, Ar cannot be liquified by increasing the pressure. **(Liquid properties)**

26. a. network (macromolecular) solid, b. metallic solid, c. ionic solid, d. molecular solid **(Solid properties)**

27. One-eighth of the eight corner atoms are inside of the unit cell, yielding one net atom. One-half of the six atoms in the faces are inside of the unit cell, yielding three atoms. Hence, the net total atoms in a face-centered unit cell is four. **(Structures of solids)**

28. 100 g $H_2O \times 4.184$ J/(g°C) $\times 100°\text{C} \times (1\text{ kJ}/1000\text{ J}) = 41.8$ kJ

    100 g $H_2O \times (1\text{ mol } H_2O/18.0\text{ g}) \times (40.8\text{ kJ/mol}) = 227$ kJ

    41.8 kJ + 227 kJ = 269 kJ **(Liquid properties)**

29. $V = a^3 = (409\text{ pm})^3 \times (1\text{ m}/10^{12}\text{ pm})^3 \times (10^2\text{ cm}/1\text{ m})^3 = 6.84 \times 10^{-23}\text{ cm}^3$

    Each unit cell of a face-centered cube has four atoms.

    4 atoms Ag $\times (1\text{ mol Ag}/6.02 \times 10^{23}\text{ atoms}) \times (108\text{ g Ag/mol}) = 7.17 \times 10^{-22}$ g Ag

    $d = \text{mass/volume} = 7.17 \times 10^{-22}\text{ g}/6.84 \times 10^{-23}\text{ cm}^3 = 10.5\text{ g/cm}^3$ **(Structure of solids)**

30. b. Segment BC of the curve shows when the solid and liquid state of Q is in equilibrium. **(Changes of state)**

31. c. Segment CD shows the increase in temperature that occurs when liquid Q is heated. **(Changes of state)**

32. a. **(Changes of state)**

33. d. The enthalpy of vaporization is the amount of heat needed to change a liquid, D, to a vapor, E, at the boiling point. **(Changes of state)**

34. a. Heat added to a solid at its melting point increases the potential energy of the particles but not the kinetic energy. **(Changes of state)**

35. b. $NH_3(l)$ is the hydrogen bonded liquid because it has an H atom bonded to the highly electronegative N atom. **(Intermolecular forces)**

36. c. All of these molecules have trigonal pyramidal geometry and to some degree are polar, but the boiling point depend mainly on the strength of the London forces. $PBr_3(l)$ has the highest boiling point because it has the strongest London forces due to having the greatest number of polarizable electrons. **(Intermolecular forces)**

37. a. $C_5H_{12}(l)$ has the lowest molar enthalpy of vaporization because it has the weakest London forces as a result of having the smallest number of electrons. **(Intermolecular forces)**

38. d. ln(1.00 atm/0.395 atm) = (ΔH_{vap}/0.008314 kJ/mol K) × ((350 K − 324 K)/350 K 324 K)

    ΔH_{vap} = 33.7 kJ/mol (**Liquid properties**)

39. c. $SiO_2(s)$ has the highest melting point because it is a covalent (network) solid. (**Solid properties**)

40. b. Substance X is in the liquid phase. (**Phase diagram**)

41. c. Substance X is in the gas phase. (**Phase diagram**)

42. e. Both a and b are correct because line xy is the point when the solid and liquid phases of substance X are in equilibrium. This is the point when it melts or freezes. (**Phase diagram**)

43. c. It boils at 75°C because line xz is the vapor pressure curve. Boiling occurs when the vapor pressure of the liquid equals 1.0 atm. (**Phase diagram**)

44. a. At 0.1 atm the vapor condenses to a liquid and at 1.0 atm it freezes. (**Phase diagram**)

45. e. The triple point conditions are 0.1 atm and −10°C. (**Phase diagram**)

Grade Yourself

Circle the numbers of the questions you missed, then fill in the total incorrect for each topic. If you answered more than three questions incorrectly, you need to focus on that topic. (If a topic has less than three questions and you had at least one wrong, we suggest you study that topic also. Read your textbook, a review book, or ask your teacher for help.)

Subject: Chemistry Fundamentals

Topic	Question Numbers	Number Incorrect
Intermolecular forces	1, 2, 3, 4, 5, 6, 35, 36, 37,	
Liquid properties	7, 8, 9, 10, 23, 24, 25, 28, 38	
Solid properties	11, 12, 26, 39	
Structure of solids	13, 14, 15, 16, 17, 18, 19, 20, 27, 29,	
Changes of state	21, 22, 30, 31, 32, 33, 34,	
Phase diagram	40, 41, 42, 43, 44, 45	

Solutions and Colloids

 Brief Yourself

Solutions are homogeneous mixtures of substances composed of one or more solutes and a solvent. The solute is usually the component present in smaller amount, and the solvent is the one present in larger amount. Mutually soluble liquids are termed miscible liquids.

Molarity is the moles of solute per liter of solution. Mole fraction is the ratio of the moles of a component to the total number of moles of all of components in the solution. Percent by mass is the mass of solute per 100 g of solution. Molality is the moles of solute per kilogram of solvent. Solutions that have a low concentration of solute are called dilute solutions and those with a high concentration are termed concentrated solutions.

Dissolution is the term used to describe the incorporation of a solute into a solvent. The dissolution of a solute is a spontaneous process that depends on the energy and randomness factors. When a solute dissolves in a solvent, the attractive forces among the solute particles break and the solvent molecules surround and attract the solute particles in a process called solvation. Overcoming the forces of attraction in the solute and solvent (bond breaking) is an endothermic process, and the solvation of the solute particles (bond making) is an exothermic process. The overall energy transferred when a solute enters solution is the enthalpy of solution, ΔH_{soln}.

A solution with the maximum amount of dissolved solute is a saturated solution. Before the point of saturation the solution is unsaturated. In a saturated solution, the dissolved solute is in equilibrium with the undissolved solute. The amount of a given solute required to just saturate a fixed amount of solvent at constant temperature is called solubility. Solubility depends on the nature of the solute and solvent molecules, temperature, and pressure. Generally, likes dissolve likes—solutes and solvents with similar structures and intermolecular forces tend to be soluble. For most solutes an increase in temperature increases the solubility. Pressure mainly affects gaseous solutions. As the partial pressure of the gas over the solution increases, its solubility increases (Henry's law).

When ionic compounds dissolve in water they dissociate and produce ions in solutions. Some polar covalent substances ionize in solutions; e.g., HCl and $HClO_4$. Solutes that produce ions in solution are termed electrolytes and those that do not are called nonelectrolytes.

Colligative properties of solutions are interrelated properties that depend on the concentration of the dissolved particles. When a nonvolatile solute dissolves in a solvent, it lowers the vapor pressure. Solutions of nonvolatile electrolytes always have lower vapor pressures than the pure solvent solution. Osmosis is the net movement of solvent molecules across a semipermeable membrane (osmotic membrane) from a weaker solution to a stronger solution.

Colloids are homogeneous mixtures with a dispersed phase in a dispersing medium. Particles dispersed in colloidal mixtures are larger than those in solutions, ranging in size range from 1 to 200 nm. Sols, gels, emulsions, foams, and aerosols are the principal colloids. Particles in the dispersed phase of a colloid are large enough to scatter light (Tyndall effect). Aqueous colloids are classified as being either hydrophilic or hydrophobic colloids.

Test Yourself

1. a. What is a solution? b. Give three examples of solutions.

2. Calculate the mole fraction of KNO_3 in a 10.0% (m/m) aqueous KNO_3 solution.

3. Calculate the mole fraction of HNO_3 in a 15.9 M HNO_3 solution. The density of this solution is 1.42 g/cm^3.

4. An acetone solution contains 4.30 mg of acetone in enough water to have a total volume of 10.7 L. What is the concentration of acetone in ppm? The density of water is 0.997 g/mL.

5. A solution contains 8.95 g KBr dissolved in 78.3 cm^3 of water. If the density of water is 0.997 g/cm^3, calculate the molality of the solution.

For Problems 6 to 9, consider an ethanol solution that was prepared by mixing 25.0 cm^3 of ethanol, C_2H_6O, with 25.0 cm^3 of H_2O. The densities of ethanol and water are 0.789 g/cm^3 and 0.997 g/cm^3, respectively. (Assume that the volumes of the liquids are additive.)

6. What is the molality of the ethanol solution?

7. What is the mole fraction of ethanol in the solution?

8. What is the molarity of the ethanol solution?

9. Consider the dissolution of sodium bromide, NaBr(s).

$$NaBr(s) \xrightarrow{H_2O} Na^+ (aq) + Br^- (aq)$$

The lattice energy of NaBr is 728 kJ/mol and its solvation energy is –741 kJ/mol. Estimate the molar enthalpy of solution for NaBr(s) and compare it to the experimentally determined value of –0.6 kJ/mol

10. Explain why ethanol, CH_3CH_2OH, and carbon tetrachloride, CCl_4, are immiscible liquids, but ethanol and water are miscible liquids.

11. At sea level the partial pressure of O_2 is 159 torr. a. If the Henry's law constant for O_2 is 1.3×10^{-3} M/atm at 25°C, calculate molar concentration of O_2 in water at 25°C. b. What is the concentration of O_2 in g/100 mL H_2O at 25°C?

12. Write an equation that shows the electrolyte behavior of $MgCl_2$.

13. A solution contains 35.5 g of glucose, $C_6H_{12}O_6$, in 95.5 g of water. If the vapor pressure of pure water is 23.8 torr at 25°C, what is the vapor pressure of the solution?

14. Urea is a water-soluble nonvolatile-nonelectrolyte solute. Calculate the mass of urea, CH_4N_2O, that must be dissolved in 275 g H_2O at 25°C to lower its vapor pressure by 2.00 torr. The vapor pressure of pure water is 23.8 torr at 25°C (298 K).

15. What is the boiling point of a solution that has 43.1 g of ethylene glycol, $C_2H_6O_2$, (a nonvolatile-nonelectrolyte) dissolved in 123 g H_2O? The K_b value for water is 0.51°C/m.

16. A nonvolatile-nonelectrolyte solute depresses the freezing point of benzene to 4.90°C. The freezing point and molal-freezing-point depression constant for benzene are 5.53°C and 5.12 °C/m, respectively. Calculate the molality of the benzene solution.

17. A 1.25-g sample of an unknown organic compound is dissolved in 53.7 g of cyclohexane, C_6H_{12}, and the solution freezes at 1.32°C. If cyclohexane freezes at 6.54°C and its molal-freezing-point-depression constant is 20.0°C/m, calculate the molar mass of the unknown compound.

18. The osmotic pressure of a solution that has 6.74 g of an unknown compound in 1.00 L of solution is 3.58 torr at 298 K. Calculate the molar mass of the unknown compound.

19. A solution contains 74.6 mg KCl dissolved in 100 g H_2O. The freezing point depression of this solution is 0.0361°C. Calculate the value for the van't Hoff factor, i.

20. A 50.0% (m/m) aqueous ethanol, CH_3CH_2OH, solution has a density of 0.9139 g/cm^3. What is the molality of the solution?

21. What is the molar concentration of NO_3^- in a solution that contains 1.00 g $Ca(NO_3)_2$ dissolved in 1.00 L of solution?

22. A saturated solution of silver arsenate, Ag_3AsO_4, has 8.5×10^{-4} g silver arsenate per 100 g H_2O. Calculate the concentration of silver arsenate in ppm.

23. Explain specifically how $KNO_3(s)$ dissolves in water.

24. The lattice and hydration energies for AgCl are 916 and –851 kJ/mol, respectively. Calculate the enthalpy of solution of AgCl.

25. If you were given a beaker that contains a colorless saturated solution and one with a colorless unsaturated solution with the same solute, explain how you would distinguish one from the other.

26. Acetic acid, CH_3COOH ($HC_2H_3O_2$), is miscible with water, but hexanoic acid, $C_5H_{11}COOH$, is only slightly soluble in water. Explain the difference in solubility of these two acids which both form hydrogen bonds with the water.

27. At 273 K and 1.00 atm N_2, 0.028 g N_2 dissolves per liter of water. What pressure of N_2 is needed to dissolve 0.10 g N_2/L H_2O at 273 K?

28. What is the freezing point of a solution that has 6.33 g $C_6H_{12}O_6$ dissolved in 88.9 g H_2O. The molal freezing point depression constant for water is 1.86°C/m.

29. A 12.47-g sample of an unknown C-H compound is dissolved in 62.11 g of benzene, C_6H_6, and the resulting solution boils at 83.40°C. The percent composition of the compound is 93.4% C and 6.6% H. Calculate the molecular formula of the compound. The normal boiling point and K_b constant for benzene are 80.1°C and 2.53 °C/m, respectively.

30. The van't Hoff factors for 0.001 M NaCl and 0.001 M MgSO$_4$ are 1.87 and 1.21, respectively. Explain why the value of the van't Hoff factor of MgSO$_4$ is significantly lower than that of NaCl.

31. What is the difference between a hydrophilic and hydrophobic colloid. What is the meaning of hydrophilic and hydrophobic?

32. An ideal solution results when you mix the two volatile liquids benzene, C_6H_6, and octane, C_8H_{18}. At 60°C the vapor pressure of benzene and octane are 385 torr and 78.3 torr, respectively. Calculate the vapor pressure of a solution prepared by mixing equal number of moles of benzene and octane.

33. One liter of an aqueous solution contains 20.0 g hemoglobin. This solution has an osmotic pressure of 6.0 torr at 25°C. Calculate the molar mass of hemoglobin.

34. What is the vapor pressure of a solution that has 6.65 g $C_6H_{12}O_6$ in 41.8 g H_2O at 14°C. The vapor pressure of water at 14°C is 11.99 torr.

35. In which of the following solvents would you predict that $I_2(s)$ is most soluble?

    a. $CCl_4(l)$

    b. $CH_3CH_2OH(l)$

    c. $H_2O(l)$

    d. $NH_3(l)$

    e. insoluble in all of these solvents

36. Calculate the mole fraction of H_2O in a 25.0% (m/m) aqueous NaOH solution.

    a. 0.870

    b. 0.750

    c. 0.130

    d. 0.417

    e. none of these

37. A solution contains 0.100 g of table sugar, $C_{12}H_{22}O_{11}$, in 12.50 L of solution. If the density of water is 0.9970 g/cm³, calculate the concentration of sugar in ppm.

    a. 8.02 ppm

    b. 802 ppm

    c. 8.00 ppm

    d. 800 ppm

    e. none of these

38. An aqueous solution has 12.1 g $NaNO_3$ dissolved in 145.9 g of solution. What is the molality of $NaNO_3$?

    a. 1.06 m

    b. 0.973 m

    c. 0.0829 m

    d. 0.142 m

    e. none of these

39. What is the molality of the solution prepared by mixing 1.00 cm³ of methanol, CH_3OH, with 99.00 cm³ of H_2O. The densities of methanol and water are 0.791 g/cm³ and 0.997 g/cm³, respectively. Assume that the volumes of the liquids are additive.

    a. 0.100 m

    b. 0.247 m

    c. 8.01 m

    d. 0.250 m

    e. none of these

40. In which of the following liquids is $CHCl_3$ miscible?

    a. CH_3COOH

    b. CH_3CH_2OH

    c. H_2O

    d. CCl_4

    e. all are miscible

Check Yourself

1. a. A solution is a homogeneous mixture of pure substances. b. Sugar water, air, $CO_2(aq)$ **(Solutions)**

2. g H_2O = 100.0 g soln – 10.0 g KNO_3 = 90.0 g

    mol KNO_3 = 10.0 g KNO_3 × (1 mol KNO_3/101 g KNO_3) = 0.0990 mol KNO_3

    mol H_2O = 90.0 g H_2O × (1 mol H_2O/18.0 g H_2O) = 5.00 mol H_2O

    X_{KNO_3} = mol KNO_3/mol total = 0.0989 mol KNO_3/5.10 mol = 0.0194 **(Concentration)**

3. mass of solution = 1 L × (1000 mL/L) × (1.42 g/mL) = 1.42 × 10³ g soln

    mass of water = 1.42 × 10³ g soln – 1.00 × 10³ g HNO_3 = 4.2 × 10² g H_2O

    moles of water = 4.2 × 10² g H_2O × (1 mol H_2O/18.0 g H_2O) = 23 mol H_2O

    X_{HNO_3} = 15.9 mol HNO_3/(15.9 mol HNO_3 + 23 mol H_2O) = 0.41 **(Concentration)**

4. 4.30 mg acetone × (1 g acetone/1000 mg acetone) = 4.30 × 10⁻³ g acetone

    10.7 L H_2O × 1000 mL/L × (0.997 g H_2O/mL H_2O) = 1.07 × 10⁴ g H_2O

    ppm acetone = (4.30 × 10⁻³ g acetone/1.07 × 10⁴ g H_2O) × 10⁶ = 0.402 ppm **(Concentration)**

5. mol KBr = 8.95 g KBr × 1 mol KBr/119 g KBr = 0.0752 mol KBr

    mass H_2O = 78.3 cm^3 H_2O × (0.997 g H_2O/1 cm^3 g H_2O) × (kg H_2O/1000 g H_2) = 0.0781 kg H_2O

    m = mol KBr/kg H_2O = 0.0752 mol KBr/0.0781 kg H_2O = 0.963 m (**Concentration**)

6. mol C_2H_6O = 25.0 cm^3 C_2H_6O × (0.789 g C_2H_6O/1 cm^3 C_2H_6O) × (1 mol C_2H_6O/46.0 g C_2H_6O) = 0.429 mol C_2H_6O

    kg H_2O = 25.0 cm^3 H_2O × (0.997 g H_2O/1 cm^3 H_2O) × (1 kg H_2O/1000 g H_2O) = 0.0249 kg H_2O

    m = 0.429 mol C_2H_6O/0.0249 kg H_2O = 17.2 m C_2H_6O (**Concentration**)

7. mol of H_2O = 0.0249 kg H_2O × (1000 g H_2O/kg H_2O) × (1 mol H_2O/18.0 g H_2O) = 1.38 mol H_2O

    $X_{C_2H_6O}$ = 0.429 mol C_2H_6O/(0.429 mol C_2H_6O + 1.38 mol H_2O) = 0.237 (**Concentration**)

8. $M_{C_2H_6O}$ = 0.429 mol C_2H_6O/0.0500 L = 8.58 M C_2H_6O (**Concentration**)

9. $NaBr(s) \rightarrow Na^+(g) + Br^-(g)$ $\Delta H_{lattice}$ = 728 kJ

    $Na^+(g) + Br^-(g) \rightarrow Na^+(aq) + Br^-(aq)$ $\Delta H_{solvation}$ = –741 kJ

    ___

    $NaBr(s) \rightarrow Na^+(aq) + Br^-(aq)$ ΔH_{soln} = –13 kJ

    This resulting value, –13 kJ, tells us that NaBr dissolves exothermically. The calculated value is not in good agreement with the experimentally obtained value of –0.6 kJ. (**Solution process**)

10. Ethanol molecules have strong hydrogen bonds and carbon tetrachloride molecules have rather weak London dispersion forces. Thus, more energy is needed to separate the hydrogen bonds among CH_3OH molecules than is released when CH_3OH and CCl_4 molecules bond because of the weak attractive forces between them. The opposite is true when considering the miscibility of water and ethanol. Both are hydrogen-bonded liquids that require a significant amount of energy to separate, but this energy is recovered when the CH_3OH and H_2O molecules hydrogen bond with each other. (**Solubility**)

11. a. $C_g = kP_g$ = 1.3 × 10^{-3} M/atm × (159 torr × 1 atm/760 torr) = 2.7 × 10^{-4} M O_2

    b. 100 mL × (1 L/1000 mL) × (2.7 × 10^{-4} mol O_2/L) × (32 g O_2/Mol O_2) = 8.6 x 10^{-4} g O_2/100 mL H_2O (**Henry's law**)

12. $MgCl_2(s) \rightarrow Mg^{2+}(aq) + 2Cl^-(aq)$ (**Electrolytes**)

13. mol H_2O = 95.5 g H_2O × (1 mol H_2O/18.0 g H_2O) = 5.31 mol H_2O

    mol $C_6H_{12}O_6$ = 35.5 g $C_6H_{12}O_6$ × (1 mol $C_6H_{12}O_6$/1.80 × 10^2 g $C_6H_{12}O_6$) = 0.197 mol $C_6H_{12}O_6$

    X_{H_2O} = mol H_2O/(mol H_2O + mol $C_6H_{12}O_6$) = 5.31 mol H_2O/(5.31 mol H_2O + 0.197 mol $C_6H_{12}O_6$) = 0.964

    $P = X_{H_2O}P_{H_2O}$ = 0.964 × 23.8 torr = 22.9 torr (**Colligative properties**)

14. $\Delta P = X_{CH_4N_2O}P_{H_2O}$

    2.00 torr = $X_{CH_4N_2O}$ × 23.8 torr

    $X_{CH_4N_2O}$ = 2.00 torr/23.8 torr = 0.0840

    mol H_2O = 275 g H_2O × (1 mol H_2O/18.0 g H_2O) = 15.3 mol H_2O

    $X_{CH_4N_2O}$ = mol CH_4N_2O/(mol CH_4N_2O + mol H_2O)

    0.0840 = mol CH_4N_2O/(mol CH_4N_2O + 15.3 mol H_2O)

    Let x equal the number of moles of CH_4N_2O.

    0.0840 = x/(x + 15.3)

    x = 1.41 mol CH_4N_2O

96 / College Chemistry

1.41 mol CH_4N_2O × (60.1 g CH_4N_2O/1 mol CH_4N_2O) = 84.7 g CH_4N_2O **(Colligative properties)**

15. 43.1 g $C_2H_6O_2$ × (1 mol $C_2H_6O_2$/62.0 g $C_2H_6O_2$) = 0.695 mol $C_2H_6O_2$

    123 g H_2O × (1 kg H_2O/1000 g) = 0.123 kg H_2O

    m = mol $C_2H_6O_2$/kg H_2O = 5.65 m $C_2H_6O_2$

    $\Delta T_b = K_b m$ = 0.51°C/m × 5.65 m = 2.9°C

    bp = 100.0°C + 2.9°C = 102.9°C **(Colligative properties)**

16. ΔT_f = 5.53°C − 4.90°C = 0.63°C

    $\Delta T_f = K_f m$

    m = $\Delta T_f/K_f$ = 0.63 °C/5.12 °C/m = 0.12 m **(Colligative properties)**

17. ΔT_f = 6.54°C − 1.32°C = 5.22°C

    $\Delta T_f = K_f m$

    m = $\Delta T_f/K_f$ = 5.22 °C/20.0 °C/m = 0.261 mol/kg C_6H_{12}

    mol of solute = 0.261 mol solute/kg C_6H_{12} × 53.7 g C_6H_{12} × (1 kg C_6H_{12}/1000 g C_6H_{12}) = 0.0140 mol solute

    Molar mass = 1.25 g solute/0.0140 mol solute = 89.2 g solute/mol solute **(Colligative properties)**

18. π = 3.58 torr × (1 atm/760 torr) = 4.71 × 10^{-3} atm

    π = MRT

    M = π/RT = 4.71 × 10^{-3} atm/(0.0821 L atm/mol K × 298 K) = 1.93 × 10^{-4} mol/L

    molar mass = g compound/mol = 6.74 g/(1.93 × 10^{-4} mol/L × 1.00 L) = 3.50 × 10^4 g/mol

    (Colligative properties)

19. 74.6 mg × (1 g/1000 mg) × (1 mol KCl/74.6 g KCl) = 0.00100 mol KCl

    100 g H_2O × 1 kg H_2O/1000 g H_2O = 0.100 kg H_2O

    m = mol KCl/kg H_2O = 0.00100 mol KCl/0.100 kg H_2O = 0.0100 m KCl

    $\Delta T_f = iK_f m$

    i = $\Delta T_f/K_f m$ = 0.0361 °C/(1.86°C/m 0.0100 m) = 1.94 **(Electrolyte solutions)**

20. 50.0% CH_3CH_2OH = 50.0 g CH_3CH_2OH/100 g soln

    50.0 g CH_3CH_2OH × (1 mol CH_3CH_2OH/46.0 g) = 1.09 mol CH_3CH_2OH

    100 g soln × (1 cm³/0.9139 g) × (1 L/1000 cm³) = 0.109 L

    M = 1.09 mol/0.109 L = 10.0 M CH_3CH_2OH **(Concentration)**

21. 1.00 g $Ca(NO_3)_2$ × (1 mol $Ca(NO_3)_2$/164.1 g) × (2 mol NO_3^-/1 mol $Ca(NO_3)_2$) = 0.0122 mol NO_3^-

    M = 0.0122 mol NO_3^-/1.00 L = 0.0122 M NO_3^- **(Concentration)**

22. ppm = (8.5 × 10^{-4} g As_3AsO_4/100 g total) × 10^6 = 8.5 ppm As_3AsO_4 **(Concentration)**

23. The ionic bonds in the KNO_3 crystal must be broken (lattice energy) and some of the water molecules' hydrogen bonds must be broken. The K^+ and NO_3^- ions are then surrounded and bonded to water molecules (hydration energy). **(Solution process)**

24. $\Delta H_{soln} = \Delta H_{lattice} + \Delta H_{hydration}$ = 916 kJ + (−851 kJ) = 65 kJ **(Solution process)**

25. Add a small amount of the solute to each solution. In the saturated solution, the solute will not dissolve, but in the unsaturated solution, it will dissolve. **(Solution properties)**

26. The hexanoic acid molecule is quite nonpolar due to the C_5H_{11} nonpolar chain even though it has a polar COOH group. Hence, water molecules cannot hydrate hexanoic acid molecules to the degree to which they hydrate the smaller two C chain in acetic acid. (**Solubility**)

27. $P_1/C_1 = P_2/C_2$

    $P_2 = P_1/C_1 \times C_2 = (1.00 \text{ atm}/0.028 \text{ g}) \times 0.10 \text{ g} = 3.6 \text{ atm}$ (**Henry's law**)

28. 6.33 g $C_6H_{12}O_6 \times$ (1 mol $C_6H_{12}O_6$/180 g) = 0.0352 mol $C_6H_{12}O_6$

    88.9 g $H_2O \times$ (1 kg/1000 g) = 0.0889 kg

    m = 0.0352 mol $C_6H_{12}O_6$/0.0889 kg H_2O = 0.396 m

    ΔT_f = 0.396 $m \times 1.86°C/m$ = 0.736°C

    fp = 0.000°C − 0.736°C = −0.736°C (**Colligative properties**)

29. ΔT_b = 83.40°C − 80.1°C = 3.3°C

    $\Delta T_b = K_b m$

    $m = \Delta T_b/K_b$ = 3.3 °C/2.53 °C/m = 1.3 mol/kg

    1.3 mol solute/kg × 0.06211 kg = 0.0810 mol

    Molar mass = 12.47 g/0.0810 mol = 154 g/mol

    93.4 g C × (1 mol C/12 g) = 7.8 mol C/6.6 mol = 1.2 × 5 = 6

    6.6 g C × (1 mol H/1.0 g) = 6.6 mol H/6.6 mol = 1 × 5 = 5

    Empirical formula C_6H_5, Empirical formula mass = 77

    Molar mass/empirical formula mass = 2

    $C_6H_5 \times 2 = C_{12}H_{10}$ (**Colligative properties**)

30. The van't Hoff factor essentially gives the effective number of dissolved particles. Because the magnesium and sulfate ions have 2+ and 2− charges, respectively, they associate to a greater degree than the 1+ and 1− ions in NaCl. Greater association of ions produces a lower van't Hoff factor. (**Electrolyte solutions**)

31. Hydrophilic (water-loving) colloids have water as the dispersing medium. Hydrophobic (water-hating) colloids can only exist in water if it stabilized in some way. (**Colloids**)

32. $P_{benzene} = X_{benzene} \times P°_{benzene}$ = 0.50 × 386 torr = 193 torr

    $P_{octane} = X_{octane} \times P°_{octane}$ = 0.50 × 78.3 torr = 39.2 torr

    P_{total} = 193 torr + 39.2 torr = 232 torr (**Colligative properties**)

33. π = 6.0 torr × (1 atm/760 torr) = 7.9 × 10^{-3} atm

    $\pi = MRT$

    $M = \pi/RT$ = 7.9 × 10^{-3} atm/(0.0821 L atm/mol K × 298 K) = 3.2 × 10^{-4} mol/L

    molar mass = g compound/mol = 20.0 g/(3.2 × 10^{-4} mol/L × 1.00 L) = 6.2 × 10^4 g/mol (**Colligative properties**)

34. 41.8 g H$_2$O × (1 mol H$_2$O/18.0 g H$_2$O) = 2.32 mol H$_2$O

6.65 g C$_6$H$_{12}$O$_6$ × (1 mol C$_6$H$_{12}$O$_6$/1.80 × 10^2 g C$_6$H$_{12}$O$_6$) = 0.0369 mol C$_6$H$_{12}$O$_6$

X_{H_2O} = mol H$_2$O/(mol H$_2$O + mol C$_6$H$_{12}$O$_6$) = 2.32 mol H$_2$O/2.36 mol total = 0.983

$P = X_{H_2O}P_{H_2O}$ = 0.983 × 11.99 torr = 11.8 torr (**Colligative properties**)

35. a. CCl$_4$(*l*) is the only nonpolar solvent listed. It is the only one that would effectively dissolve the nonpolar I$_2$ molecules. (**Solubility**)

36. a. 75.0 g H$_2$O × (1 mol H$_2$O/18.0 g H$_2$O) = 4.17 mol H$_2$O

    25.0 g NaOH × (1 mol NaOH/40.0 g) = 0.625 mol NaOH

    X_{H_2O} = mol H$_2$O/mol total = 4.17 mol H$_2$O/4.79 mol total = 0.870   (**Concentration**)

37. a. 12.50 L × (1000 cm^3/1 L) × (0.9970 g/cm^3) = 1.246 × 10^4 g H$_2$O

    ppm = (0.100 g/1.246 × 10^4 g H$_2$O) × 10^6 = 8.02 ppm (**Concentration**)

38. a. 145.9 g soln − 12.1 g NaNO$_3$ = 133.8 g H$_2$O × (1 kg/1000 g) = 0.1338 kg H$_2$O

    12.1 g NaNO$_3$ × 1 mol NaNO$_3$/85.0 g = 0.142 mol NaNO$_3$

    m = 0.142 mol NaNO$_3$/0.1338 kg = 1.06 *m* NaNO$_3$ (**Concentration**)

39. d. 1.00 cm^3 × (0.791 g/cm^3) × (1 mol CH$_3$OH/32.0 g) = 0.0247 mol CH$_3$OH

    99.00 cm^3 × (0.997 g/cm^3) × (1 kg H$_2$O/1000 g) = 0.0987 kg H$_2$O

    m = 0.0247 mol/0.0987 kg = 0.250 *m* (**Concentration**)

40. d. CCl$_4$ and CHCl$_3$ are miscible liquids because their molecules are similar in size and intermolecular forces. (**Solubility**)

Grade Yourself

Circle the numbers of the questions you missed, then fill in the total incorrect for each topic. If you answered more than three questions incorrectly, you need to focus on that topic. (If a topic has less than three questions and you had at least one wrong, we suggest you study that topic also. Read your textbook, a review book, or ask your teacher for help.)

Subject: Solutions and Colloids

Topic	Question Numbers	Number Incorrect
Solutions	1	
Concentration	2, 3, 4, 5, 6, 7, 8, 20, 21, 22, 36, 37, 38, 39	
Solution process	9, 23, 24	
Solution properties	25	
Solubility	10, 26, 35, 40	
Henry's law	11, 27	
Electrolytes and electrolyte solutions	12, 19, 30	
Colligative properties	13, 14, 15, 16, 17, 18, 28, 29, 32, 33, 34	
Colloids	31	

Chemical Kinetics–Rates of Chemical Reactions

 Brief Yourself

Chemical kinetics is the study of the rates of chemical reactions and the underlying mechanisms by which reactants change to products. Concentration, temperature, catalysts, and surface area are the factors that affect the rates of chemical reactions.

Reaction rates are determined by measuring the change in concentration of reactants or products over a time interval. If the concentration change is measured over a relatively long time interval, the average reaction rate is obtained. If the concentration is measured over a very short interval, the instantaneous reaction rate is obtained.

To study the effect of reactant concentration on reaction rates, the initial instantaneous rate at different concentrations are measured. From such data the rate law for a reaction is determined. The rate law is an equation that shows the initial instantaneous reaction rate in terms of the concentrations of the reactants raised to a power. In addition, a rate law has the rate constant which is characteristic of a reaction at a constant temperature. The general form for the rate law is

$$\text{rate} = k[A]^x[B]^y$$

in which k is the rate constant, $[A]$ and $[B]$ are the molar concentrations of the reactants, and x and y are the reactions orders—the exponents of the concentration terms. An exponent of 1 means that it is first order with respect to that reactant, and an exponent of 2 means it is second order. Reaction orders may be zero or fractional and the sum the reaction orders is the overall reaction order.

The relationship of time versus concentration for a first-order reaction is

$$\ln([A]_0/[A]_t) = kt$$

This equation may be rearranged and expressed as a linear relationship between the $\ln[A]_t$ and t in which k is the slope and $\ln[A]_0$ is the y intercept.

$$\ln[A]_t = -kt + \ln[A]_0$$

It is convenient to measure the half-lives for reactions. The half-life is the time for one-half of the reactant molecules to change to products ($[A]_t = \frac{1}{2}[A]_0$). For first-order reactions, the half-life is calculated from the following equation.

$$t_{1/2} = 0.693/k \quad \text{(first order)}$$

The following equation is used to investigate time versus concentration for reactions that follow second-order kinetics with respect to a single reactant.

$$1/[A]_t = kt + 1/[A]_0$$

Second-order equations of this type exhibit a linear relationship between $1/[A]_t$ and t, and the slope of the line is k and the y intercept is $1/[A]_0$. Second-order half lives are calculated by taking the reciprocal of the product of the rate constant and initial concentration

$$t_{1/2} = 1/(k[A]_0)$$

Collision theory explains the rates of reactions in terms of collisions among the reacting molecules. The rate of a reaction is proportional to the number of collisions per second between the reacting molecules and the fraction of effective collisions. An effective collision is one that yields the products. Such collisions require that the reactant particles have the proper orientation and an energy of collision equal to or greater than the activation energy, E_a. The activation energy is the minimum energy to produce the products. When the activation energy is added to reacting particles with the correct orientation the activated complex (transition state) results. An activated complex is a high-energy transient species formed from colliding molecules in which reactant bonds are partially broken and product bonds are partially formed.

A temperature rise increases the average kinetic energy of the reactant particles and allows more of them to reach the activated complex. Thus, when the temperature increases, the reaction rate increases. The Arrhenius equation is used to calculate the rate constants at different temperatures. One way to express the Arrhenius equation is as follows.

$$\ln k = -E_a/RT + \ln A$$

A reaction mechanism is a description of the series of steps that occur as the reactants change to products. Reaction mechanisms describe how reactions occur. They can be elucidated experimentally by collecting rate data. The elementary reactions are the individual steps in a reaction mechanism. The number of reactants that produce the transition state in an elementary reaction is the molecularity. If a single molecule or ion produces one or more molecules in an elementary reaction, then it is classified as being unimolecular. When two particles collide in an elementary step it is a bimolecular reaction, and if three particles collide it is a termolecular reaction. In a reaction mechanism, the slowest step usually determines the overall rate of the reaction and is known as the rate-determining step.

Catalysts are substances that increase the rates of chemical reactions and may be recovered unchanged after the reaction. Homogeneous catalysts are in the same phase as the reactants, and heterogeneous catalysts are in a different phase than the reactants. Catalysts speed up reactions by lowering the activation energies of reactions.

Test Yourself

1. Consider the following equation:

   $$F_2(g) + 2NO(g) \rightarrow 2ONF(g)$$

   Write a mathematical expression that shows the relationship of the disappearance of the reactants and the appearance of the products.

2. Consider the following gas-phase reaction:

   $$N_2(g) + 3H_2(g) \rightarrow 2NH_3(g)$$

   Show the relationships between the rates of disappearance of N_2 and H_2 and the rate of appearance of NH_3.

3. Consider the following gas-phase decomposition of dinitrogen pentoxide, N_2O_5:

$$2N_2O_5(g) \rightarrow 4NO(g) + O_2(g)$$

Write the rate expression or disappearance of reactant and appearance of products in terms of changes in partial pressures.

4. At a specific set of reaction conditions the rate of decomposition of $BrF_5(g)$ is -6.4×10^{-3} atm/s.

$$2BrF_5(g) \rightarrow Br_2(g) + 5F_2(g)$$

Calculate the rates at which Br_2 and F_2 form.

5. Dinitrogen pentoxide, N_2O_5, undergoes a decomposition reaction at 45°C. If the initial concentration of N_2O_5 is 0.0176 M and after 600 s elapse the concentration decreases to 0.0124 M. Calculate the average rate of the reaction over the first 600 s.

6. Consider the following equation:

$$Cl_2(g) + 2NO(g) \rightarrow 2NOCl(g)$$

The rate equation for this reaction is as follows.

Initial rate = $k[Cl_2][NO]^2$

The rate of this reaction is 6.87×10^{-7} M/s when the initial concentrations of Cl_2 and NO are 0.245 M and 0.175 M, respectively. Calculate the rate constant for this reaction.

7. Consider the following equation:

$$2NO_2 \rightarrow 2NO + O_2$$

If the reaction is second order with respect to NO_2, write the rate equation for this reaction.

8. For the reaction

$$2NO(g) + 2H_2(g) \rightarrow N_2(g) + 2H_2O(g)$$

the following rate data were collected at 900°C.

Experiment	[NO]	[H$_2$]	Initial rate, M/s
1	0.388	0.214	0.204
2	0.388	0.428	0.407
3	0.259	0.155	0.0488
4	0.777	0.155	0.439

a. Determine the rate law for the reaction.
b. Calculate the rate constant, k.

Use the following information for Problems 9 to 11.

The decomposition of acetone, CH_3COCH_3, at 600°C to CO and several C-H compounds follow first-order kinetics. A rate study reveals that it takes 12.3 s for the concentration of acetone to decrease from 0.100 M to 0.0900 M.

9. Calculate the rate constant for the decomposition reaction of acetone.

10. How long does it take for the initial concentration of 0.100 M acetone to decrease to 10.0% of this value?

11. What concentration of acetone remains if initially 0.230 M decomposes for 175 s?

12. The half-life for the following reaction is 112 s.

$$2Cl_2O_7(g) \rightarrow 2Cl_2(g) + 7O_2(g)$$

Calculate the pressure of Cl_2O_7 after 285 s when the initial pressure of Cl_2O_7 is 0.0553 atm.

Use the following information for Problems 13 and 14.

The decomposition of NO(g) to $N_2(g)$ and $O_2(g)$ follows second-order kinetics.

$$2NO(g) \rightarrow N_2(g) + O_2(g)$$

Over a 1500 s interval, the concentration of NO(g) decreases from 0.0050 M to 0.0044 M.

13. What is the second-order rate constant for this reaction?

14. What is the concentration of NO after a second 1500 s time interval?

15. What is the half-life of a reaction?

16. a. What is the activation energy for a chemical reaction? b. What affect does the activation energy have on the rate of a reaction?

Consider the following gas-phase reaction for Problems 17 and 18:

$$NO(g) + Cl_2(g) \rightarrow NOCl(g) + Cl(g)$$

The rate constant, k, for this reaction is 32.5 $M^{-1}s^{-1}$ at 550.0 K and 45.2 $M^{-1}s^{-1}$ at 560.0 K.

17. Calculate the activation energy in kJ/mol for the reaction.

18. Calculate the rate constant for the reaction at 540.0 K.

Consider the following mechanism to answer Problems 19 to 21:

The following is the proposed mechanism for the reaction of NO and Br_2 to produce NOBr.

$$NO(g) + Br_2(g) \underset{k_{-1}}{\overset{k_1}{\rightleftarrows}} NOBr_2(g) \quad \text{(fast)}$$

$$NOBr_2(g) + NO(g) \overset{k_2}{\longrightarrow} 2NOBr(g) \quad \text{(slow)}$$

19. What is the overall reaction?

20. What is the molecularity of the elementary reactions?

21. Show that this mechanism is consistent with the experimentally determined rate law, rate = $k[NO]^2[Br_2]$.

22. What is the difference between a homogenous and heterogeneous catalyst?

23. Explain how a catalyst increases the rate of a chemical reaction.

24. The rate constant for the disappearance of Cl_2 in the third-order reaction of Cl_2 and NO to form NOCl is 4.5 $M^{-2}s^{-1}$ at 0.0°C and is 8.0 $M^{-2}s^{-1}$ at 22°C. What is the activation energy for this reaction?

25. The first-order rate constant for the conversion of cyclopropane, C_3H_6, to propene, C_3H_6, is 9.2 s^{-1} at 1000°C. How long does it take for 99% of the cyclopropane to react?

26. Radioactive decay of elements follows first order kinetics. ^{18}F has a half-life of 109.7 s. How long would it take for the mass of ^{18}F to decay from 10.0 g to 0.010 g?

27. If I^- is treated with OCl^- in aqueous solution, Cl^- and OI^- result.

$$I^-(aq) + OCl^-(aq) \rightarrow Cl^-(aq) + OI^-(aq)$$

Consider the following rate data for the reaction:

$[I^-]_{initial}$	$[OCl^-]_{initial}$	Initial rate, M/s
8.25×10^{-4}	3.85×10^{-4}	1.92×10^3
1.65×10^{-3}	3.85×10^{-4}	3.84×10^3
8.25×10^{-4}	1.16×10^{-3}	5.76×10^3

What is the rate constant, k, for this reaction?

28. The following concentration-time data were collected for the decomposition of ammonia to N_2 and H_2.

$$2NH_3(g) \rightarrow N_2(g) + 3H_2(g)$$

Time, s	$[NH_3]$, M
0.0	1.00
1.0	0.991
2.0	0.983

Calculate the average rate of disappearance of NH_3 from 1.0 s to 2.0 s.

29. Calculate the activation energy of a reaction which doubles its rate going from 15°C to 30°C.

30. Consider the proposed mechanism for the gas-phase reaction of NO_2 and CO:

$$NO_2 + NO_2 \rightarrow NO_3 + NO \quad \text{(slow)}$$

$$NO_3 + CO \rightarrow NO_2 + CO_2 \quad \text{(fast)}$$

Write a rate law that is consistent with this mechanism. What are the reaction orders for NO_2 and CO?

31. The first-order rate constant for the decomposition of 0.945 M H_2O_2 to H_2O is 8.49×10^{-4} s^{-1} at a particular set of conditions. What concentration of H_2O_2 remains after 125 s?

32. It takes 195 s for 75.8% of the reactants to disappear in a first-order reaction. How long does it take for 99.9% of the reactants to disappear?

33. The first-order rate constant for the decomposition of a compound is 8.1×10^{-3} s^{-1}. What is the half life of this reaction?

34. What happens to the ratio of the rate constants, k_2/k_1, and thus the reaction rates, when the Kelvin temperature doubles from 200 K to 400 K for a reaction with an activation energy of 57 kJ?

35. Consider the following reaction:

$$N_2(g) + 3H_2(g) \rightarrow 2NH_3(g)$$

If the rate of disappearance of N_2 is -1.0×10^{-3} M/s, calculate the rate of disappearance of H_2.

a. -3.0×10^{-3} M H$_2$/s
b. -2.0×10^{-3} M H$_2$/s
c. -1.0×10^{-3} M H$_2$/s
d. -4.0×10^{-3} M H$_2$/s
e. none of these

36. Consider the following hypothetical reaction and rate data:

$$2A(g) + B_2(g) \rightarrow 2AB(g)$$

Experiment	[A]	[B$_2$]	Reaction rate, M/s
1	0.125	0.293	0.548
2	0.125	0.0977	0.183
3	0.377	0.185	0.346
4	0.250	0.586	1.10
5	0.198	0.185	0.346

Determine the rate law for this reaction.

a. rate = $k[A][B]$
b. rate = $k[A]^2[B]$
c. rate = $k[A][B]^2$
d. rate = $k[B]$
e. none of these

37. For the rate data presented in Problem 36, calculate the rate constant.

a. 1.87 s^{-1}
b. 15.0 s^{-1}
c. 119 s^{-1}
d. 4.38 s^{-1}
e. none of these

38. For the rate data presented in Problem 37, calculate the initial rate of the reaction when the concentration of A is 0.471 M and B_2 is 0.313 M.

a. 0.548 M/s
b. 0.585 M/s
c. 0.276 M/s
d. 0.881 M/s
e. none of these

39. Sucrose, $C_{12}H_{22}O_{11}$, decomposes to glucose and fructose. The concentration of sucrose drops from 0.200 M to 0.175 M in 2310 s. If this reaction is first order, calculate the time it takes for the initial concentration, 0.200 M, to drop 40.0%.

a. 1.59×10^4 s
b. 5.78×10^5 s
c. 8.84×10^3 s
d. 1.2×10^4 s
e. none of these

40. The first-order rate constant for the fermentation of sucrose, $C_{12}H_{22}O_{11}$, to ethanol, CH_3CH_2OH, and CO_2 is 1.9×10^{-5} s^{-1}. Calculate the half-life for this reaction.

a. 6.93×10^{-1} s
b. 2.74×10^{-5} s
c. 3.65×10^4 s
d. 5.26×10^4 s
e. none of these

Check Yourself

1. $-2(\Delta[F_2]/\Delta t) = -(\Delta[NO]/\Delta t) = (\Delta[ONF]/\Delta t)$ **(Rate expressions)**

2. $-(\Delta[N_2]/\Delta t) = -\frac{1}{3}(\Delta[H_2]/\Delta t) = \frac{1}{2}(\Delta[NH_3]/\Delta t)$ **(Rate expressions)**

3. $-\frac{1}{2}\Delta P_{N_2O_5}/\Delta t = \frac{1}{4}\Delta P_{NO}/\Delta t = \Delta P_{O_2}/\Delta t$ **(Rate expressions)**

4. $\Delta P_{Br_2}/\Delta t = -\frac{1}{2}(\Delta P_{BrF_5}/\Delta t) = -\frac{1}{2} \times -6.4 \times 10^{-3}$ atm/s $= 3.2 \times 10^{-3}$ atm/s

   $1/5(\Delta P_{F_2}/\Delta t) = -1/2(\Delta P_{BrF_5}/\Delta t)$

   $(\Delta P_{F_2}/\Delta t) = -5/2(\Delta P_{BrF_5}/\Delta t)$

   $\Delta P_{F_2}/\Delta t = -5/2 \times -6.4 \times 10^{-3}$ atm/s $= 1.6 \times 10^{-2}$ atm/s **(Rate expressions)**

5. Reaction rate $= -\Delta[N_2O_5]/\Delta t$

   Reaction rate $= -([N_2O_5]_{600\,s} - [N_2O_5]_0\,s)/(600\,s - 0\,s) = -(0.0124\,M - 0.0176\,M)/600\,s = 8.67 \times 10^{-6}$ M/s

   (Rate expressions)

6. Rate $= k\,[Cl_2][NO]^2$

   k = Initial rate/$([Cl_2][NO]^2) = 6.87 \times 10^{-7}$ M/s/$(0.245\,M \times 0.175^2\,M^2) = 9.16 \times 10^{-5}$ $1/(M^2 s)$
   (Rate expressions)

7. Initial rate $= k[NO_2]^2$ **(Rate expressions)**

8. a. Rate(Experiment 2)/Rate(Experiment 1) $= k[NO]_2 x[H_2]_2 y/k[NO]_1 x[H_2]_1 y$

   Rate(Experiment 2)/Rate(Experiment 1) $= [H_2]_2 y/[H_2]_1 y = ([H_2]_2/[H_2]_1)^y$

   $0.407\,M/s/0.204\,M/s = (0.428\,M/0.214\,M)^y$

   $2.00 = 2.00^y$

   $y = 1$

   Rate(Experiment 4)/Rate(Experiment 3) $= ([NO]_4/[NO]_3)^x$

   $0.439\,M/s/0.0488\,M/s = (0.777\,M/0.259\,M)^x$

   $9.00 = 3.00^x$

   $x = 2$

   Rate $= k[NO]^2[H_2]$

   b. Rate $= k[NO]^2[H_2]$

   k = Rate/$([NO]^2[H_2]) = 0.204\,M/s/((0.388\,M)^2 \times 0.214\,M) = 6.33\,M^{-2}s^{-1}$ **(Rate expressions)**

9. $\ln([CH_3COCH_3]_0/[CH_3COCH_3]_t) = kt$

   $\ln(0.100\,M/0.0900\,M) = k \times 12.3\,s$

   $\ln 1.11 = 12.3\,k$

   $0.105 = 12.3\,k$

   $k = 8.57 \times 10^{-3}\,s^{-1}$ **(First-order kinetics)**

10. $\ln([CH_3COCH_3]_0/[CH_3COCH_3]_t) = kt$

    $10.0\% \times 0.100\ M = 0.0100\ M$

    $\ln(0.100\ M/0.0100\ M) = 8.57 \times 10^{-3}\ s^{-1}\ t$

    $2.30 = 8.57 \times 10^{-3}\ s^{-1}\ t$

    $t = 269\ s$ **(First-order kinetics)**

11. $\ln([CH_3COCH_3]_0/[CH_3COCH_3]_t) = kt$

    $\ln(0.230\ M/[CH_3COCH_3]_{175}) = 8.57 \times 10^{-3}\ s^{-1} \times 175\ s = 1.50$

    $0.230\ M/[CH_3COCH_3]_{175} = e^{1.50} = 4.48$

    $[CH_3COCH_3]_{175} = 5.13 \times 10^{-2}\ M$ **(First-order kinetics)**

12. $t_{1/2} = 0.693/k$

    $k = 0.693/t_{1/2} = 0.693/112\ s = 6.19 \times 10^{-3}\ s^{-1}$

    $\ln(P_0/P_{285}) = kt$

    $\ln(0.0553\ atm/P_{285}) = 6.19 \times 10^{-3}\ s^{-1} \times 285\ s = 1.76$

    $0.0553\ atm/P_{285} = e^{1.76} = 5.8$

    $P_{285} = 9.5 \times 10^{-3}\ atm$ **(First-order kinetics)**

13. $1/0.0044\ M = 1/0.0050\ M + k\ 1500\ s$

    $1500\ k = 1/0.0044 - 1/0.0050 = 27.3$

    $k = 0.018\ M^{-1}\ s^{-1}$ **(Second-order kinetics)**

14. $1/[NO]_{3000} = 1/[NO]_0 + kt$

    $1/[NO]_{3000} = 1/0.0050\ M + k\ 3000\ s$

    $[NO]_{3000} = 0.0039\ M$ **(Second-order kinetics)**

15. Half-life is the amount of time it takes for one-half of the reactants to be converted to products. **(Half life)**

16. a. Activation energy, E_a, is the minimum energy required to produce the activated complex. This is the minimum energy needed to reach the transition state. b. Reactions with higher activation energies are slower reactions. **(Activation energy)**

17. $\ln(k_2/k_1) = E_a/R((T_2 - T_1)/T_1T_2)$

    $\ln(45.2\ M^{-1}s^{-1}/32.5\ M^{-1}s^{-1}) = E_a/8.314\ J/(mol\ K) \times ((560.0\ K - 550.0\ K)/(550.0\ K \times 560.0\ K))$

    $0.330 = E_a \times 3.91 \times 10^{-6}\ mol/J$

    $E_a = 8.45 \times 10^4\ J/mol \times 1\ kJ/1000\ J = 84.5\ kJ/mol$ **(Arrhenius equation)**

18. $\ln(k_2/k_1) = E_a/R((T_2 - T_1)/T_1T_2)$

    $\ln(k_2/32.5\ M^{-1}s^{-1}) = 8.45 \times 10^4\ J/mol/8.314\ J/(molK) \times ((540.0\ K - 550.0\ K)/540.0\ K \times 550.0\ K))$

    $\ln(k_2/32.5\ M^{-1}s^{-1}) = -0.342$

    $k_2/32.5\ M^{-1}s^{-1} = 0.710$

    $k_2 = 23.1\ M^{-1}s^{-1}$ **(Arrhenius equation)**

19. The overall reaction is the sum of the elementary reactions.

    NO(g) + Br$_2$(g) $\rightleftarrows$ NOBr$_2$(g)

    + NOBr$_2$(g) + NO(g) $\rightarrow$ 2NOBr(g)

    2NO(g) + Br$_2$(g) $\rightarrow$ 2NOBr(g) **(Reaction mechanism)**

20. Because both elementary reactions involve two reacting species, they are both bimolecular processes. **(Reaction mechanism)**

21. The second step is the rate-limiting step; consequently, its rate law is

    rate = k_2[NOBr$_2$][NO]

    k_1[NO][Br$_2$] = k_{-1}[NOBr$_2$]

    To remove NOBr$_2$ from the equation solve for [NOBr$_2$].

    [NOBr$_2$] = k_1/k_{-1} [NO][Br$_2$]

    Substituting this expression into the rate equation for step 2 yields the following equation:

    rate = k_2 (k_1/k_{-1} [NO][Br$_2$])[NO]

    If k equals $k_1 k_2/k_{-1}$, then the equation becomes

    rate = k[NO]^2[Br$_2$] **(Reaction mechanism)**

22. A homogeneous catalyst is in the same phase as the reactants. A heterogeneous catalyst is in a different phase than the reactants. **(Catalysts)**

23. A catalyst increases the rate of a reaction by lowering the activation energy. This means that a greater number of reacting particles have sufficient energy to produce the products at a fixed temperature. **(Catalysts)**

24. $\ln(k_2/k_1) = E_a/R((T_2 - T_1)/T_1 T_2)$

    $\ln(8.0\ M^{-2}\ s^{-1}/4.5\ M^{-2}\ s^{-1}) = E_a/0.008314\ \text{kJ/mol K} \times ((295\ K - 273\ K)/(273\ K \times 295\ K))$

    E_a = 17.5 kJ **(Arrhenius equation)**

25. 99% reacted means 1% unreacted

    $\ln [A]_0/[A]_t = kt$

    $\ln(100/1) = 9.2\ s^{-1} \times t$

    t = 0.50 s **(First order kinetics)**

26. $k = 0.693/t_{1/2} = 0.693/109.7\ s = 0.00632\ s^{-1}$

    $\ln\ ^{18}F_0/^{18}F_t = kt$

    $\ln(10.0\ g/0.010\ g) = 0.00632\ s^{-1} \times t$

    $t = 1.0 \times 10^3$ s **(First order kinetics)**

27. The reaction is first order with respect to both reactants; therefore, the rate equation is

    rate = k[I$^-$][OCl$^-$].

    rate = k (8.25 $\times$ 10^{-4} M)(3.85 $\times$ 10^{-4} M)

    $k = 6.04 \times 10^9\ M^{-1}s^{-1}$ **(Rate expressions)**

28. $-\Delta[NH_3]/\Delta t = -(0.983\ M - 0.991\ M)/(2.0\ s - 1.0\ s) = -0.008\ M/s$ (**Rate expression**)

29. $\ln(k_2/k_1) = E_a/R((T_2 - T_1)/T_1T_2)$

    $\ln(2/1) = E_a/0.008314\ kJ/mol\ K \times ((303\ K - 288\ K)/(288\ K \times 303\ K))$

    $E_a = 34\ kJ$ (**Arrhenius equation**)

30. rate = $k[NO_2]^2$ The reaction is second order with respect to NO_2 and zero with respect to CO (CO is not in the rate determining step).   (**Reaction mechanism**)

31. $\ln[H_2O_2]_0/[H_2O_2]_t = kt$

    $\ln(0.945\ M/[H_2O_2]_t) = 8.49 \times 10^{-4}\ s^{-1} \times 125\ s$

    $[H_2O_2]_t = 0.850\ M$ (**First order kinetics**)

32. $\ln[A]_0/[A]_t = kt$

    $\ln(100/24.2) = k \times 195\ s$

    $k = 0.00728\ s^{-1}$

    $\ln(100/0.100) = 0.00728\ s^{-1} \times t$

    $t = 948\ s$ (**First order kinetics**)

33. $t_{1/2} = 0.693/k = 0.693/8.1 \times 10^{-3}\ s^{-1} = 86\ s$ (**First order kinetics**)

34. $\ln(k_2/k_1) = E_a/R((T_2 - T_1)/T_1T_2)$

    $\ln(k_2/k_1) = 57\ kJ/0.008314\ kJ/mol\ K \times ((400\ K - 200\ K)/(400\ K \times 200\ K))$

    $k_2/k_1 = 2.8 \times 10^7$ (**Arrhenius equation**)

35. a. $\Delta[H_2]/\Delta t = 3 \times \Delta[N_2]/\Delta t = 3 \times -1.0 \times 10^{-3}\ M/s = -3.0 \times 10^{-3}\ M\ H_2/s$ (**Rate expressions**)

36. d. The rate is independent of [A], and is first order with respect to [B]; thus,

    rate = $k[B]$ (**Rate expressions**)

37. a. rate = $k[B]$

    $0.548\ M/s = k \times 0.293\ M$

    $k = 1.87\ s^{-1}$ (**Rate expressions**)

38. b. rate = $k[B]$

    rate = $1.87\ s^{-1} \times 0.313\ M = 0.585\ M/s$ (**Rate expressions**)

39. c. $\ln(0.200\ M/0.175\ M) = k \times 2310\ s$

    $k = 5.78 \times 10^{-5}\ s^{-1}$

    $\ln(0.200\ M/0.120\ M) = (5.78 \times 10^{-5}\ s^{-1})\ t$

    $t = 8.84 \times 10^3\ s$ (**First-order kinetics**)

40. c. $t_{1/2} = 0.693/k = 0.693/1.9 \times 10^{-5}\ s^{-1} = 3.65 \times 10^4\ s$ (**First-order kinetics**)

Grade Yourself

Circle the numbers of the questions you missed, then fill in the total incorrect for each topic. If you answered more than three questions incorrectly, you need to focus on that topic. (If a topic has less than three questions and you had at least one wrong, we suggest you study that topic also. Read your textbook, a review book, or ask your teacher for help.)

Subject: Chemical Kinetics—Rates of Chemical Reactions

Topic	Question Numbers	Number Incorrect
Rate expressions	1, 2, 3, 4, 5, 6, 7, 8, 27, 28, 35, 36, 37, 38	
First order-kinetics	9, 10, 11, 12, 25, 26, 31, 32, 33, 39, 40	
Second order-kinetics	13, 14	
Half life	15	
Activation energy	16	
Arrhenius equation	17, 18, 24, 29, 34	
Reaction mechanism	19, 20, 21, 30	
Catalysts	22, 23	

Fundamental Principles of Chemical Equilibria

 ## Brief Yourself

A chemical equilibrium establishes when the rate of the forward reaction equals the rate of the reverse reaction. At that time the concentrations of the reactants and products remain constant. The law of mass action expresses the concentration of the reactants and products in terms of an equilibrium constant, K_c. The equilibrium constant equals the product of the molar concentrations of the products each raised to the power that corresponds to their coefficients in the balanced equation divided by the product of the molar concentrations of the reactants each raised to the power corresponding to their coefficients. Pure liquids and solids are omitted from the equilibrium expression because their concentrations are constant and have no effect on the equilibrium as long as they are present. Equilibria that have reactants and products in more than one phase are heterogeneous equilibria and those in which all substances in one phase are homogeneous equilibria.

If the value of the equilibrium constant is greater than 1 ($K_c > 1$) then the magnitude of the numerator of the equilibrium expression is larger than that of the denominator. Hence, the equilibrium concentrations of the products are usually higher than those of the reactants. If the value of K_c is less than 1, the equilibrium concentrations of the reactants are generally greater than those of the products. If an equation for an equilibrium is reversed, the value of K_c becomes the reciprocal of the original constant.

Equilibrium constants for gas-phase equilibria are often expressed in terms of their partial pressures. This is known as the K_p equilibrium constant. The following equation shows the relationship of K_p to K_c.

$$K_p = K_c(RT)^{\Delta n}$$

in which R is the ideal gas constant, T is the Kelvin temperature, and Δn is the change in the number of moles of gas as determined from the coefficients of the balanced chemical equation. Equilibrium constants are obtained by measuring the equilibrium concentrations or partial pressures of each component of an equilibrium mixture and substituting these values into the equilibrium expression.

Before a system reaches equilibrium, the concentrations or partial pressures are expressed in terms of the reaction quotient, Q. The expression for Q is in the same form as that of K. The value of Q is used to determine the direction from which an equilibrium develops. If a reaction attains equilibrium from the reactants, the value of Q is less than K, and if a reaction establishes equilibrium from the products, the value of Q is greater than K. An equilibrium expression is used to calculate the equilibrium concentrations given the starting concentrations. When solving such problems always remember to sketch a graph of the concentrations or partial pressures versus time as the equilibrium establishes, and use the graph to construct a summary table that shows the initial concentrations, concentration changes, and equilibrium concentrations of all components of the equilibrium. Then substitute the equilibrium concentrations from the table into the equilibrium expression and solve for the change in concentration or pressure. Finally, calculate the equilibrium concentrations by substituting the value for the change into the equilibrium terms in the table.

Concentration, pressure, volume, and temperature changes affect the concentrations of the components of a chemical equilibrium. Their effects may be predicted by the application of Le Chatelier's principle which states that equilibrium systems tend to absorb any changes that may occur in such a way to reduce the change and return to a state of equilibrium. An increase in the concentration of a component of an equilibrium shifts in favor of the direction that consumes the added substance, and a decrease in the concentration shifts the equilibrium in favor of the direction that replaces it. The equilibrium shifts in the direction of the smaller number of moles of molecules when you increase the pressure and decrease the volume. Le Chatelier's principle predicts that an increase in temperature favors the reaction that absorbs the added heat and that a decrease in temperature favors the reaction that releases heat. When you change the temperature the equilibrium constant changes accordingly. A temperature increase decreases the value of K for an exothermic reaction and increases the value of K for an endothermic reaction. Catalysts do not affect the concentrations of chemical equilibria because they increase the rates of the forward and reverse reactions equally.

Test Yourself

1. Consider the general reaction:

   $$wA + xB \rightleftarrows yC + zD$$

   Write the equilibrium constant expression.

2. Consider the following equilibrium:

   $$H_2(g) + CO_2(g) \rightleftarrows CO(g) + H_2O(g)$$

   If a mixture of 0.100 M H_2 and 0.100 M CO_2 at 873 K is allowed to attain equilibrium, the concentrations of H_2 and CO_2 decrease to 0.0609 M and the concentrations of CO and H_2O increase to 0.0390 M. Calculate the equilibrium constant, K_c.

3. Consider the following equilibrium:

   $$CO_2(g) + C(graphite) \rightleftarrows 2CO(g)$$

   Write the K_c equilibrium expression.

4. Write the equilibrium expression for the following:

   $$3CuO(s) + 2NH_3(g) \rightleftarrows 3Cu(s) + N_2(g) + 3H_2O(g)$$

5. Consider the equilibrium which has an equilibrium constant of 0.51 at 673 K:

   $$N_2(g) + 3H_2(g) \rightleftarrows 2NH_3(g)$$

   a. Write the equilibrium expression. b. What is the equilibrium constant for the reverse reaction?

6. Consider the following equilibrium:

   $$SO_2(g) + \tfrac{1}{2}O_2(g) \rightleftarrows SO_3(g) \quad K_c = 20.4 \ (973 \ K)$$

   Determine the equilibrium constants, K_c for the following.

   $$2SO_3(g) \rightleftarrows 2SO_2(g) + O_2(g)$$

7. Consider the following gas-phase reaction:

   $$N_2O_4(g) \rightleftarrows 2NO_2(g)$$

   Write the K_p equilibrium expression for this reaction.

8. The K_c for the following equilibrium:

   $$CO(g) + Cl_2(g) \rightleftarrows COCl_2(g)$$

   is 4.6×10^9 at 373 K. Calculate the K_p for the equilibrium.

9. Consider the following equilibrium at 2773 K:

$$H_2(g) + Cl_2(g) \rightleftharpoons 2HCl(g)$$

Initially, 0.250 M H_2 and 0.250 M Cl_2 are introduced into a reaction vessel and the system is allowed to equilibrate. At equilibrium the concentrations of H_2 and Cl_2 become 0.0314 M.
a. Calculate K_c. b. Calculate K_p.

10. Consider the following gas-phase equilibrium:

$$SO_2Cl_2(g) \rightleftharpoons SO_2(g) + Cl_2(g)$$

A 3.35-g sample of SO_2Cl_2 is introduced into a 1.00-L reaction vessel and is allowed to reach equilibrium at 300 K. At equilibrium 0.120 atm SO_2 results in the vessel. What is the K_p for the equilibrium?

11. For the equilibrium in Problem 10, calculate the K_c value.

12. Calculate the equilibrium constant for the following equilibrium system at 1120 K

$$C(s) + CO_2(g) + 2Cl_2(g) \rightleftharpoons 2COCl_2(g)$$

given the following equations and equilibrium constants:

$$CO(g) + Cl_2(g) \rightleftharpoons COCl_2(g) \quad K_p(1) = 6.0 \times 10^{-3}$$

$$C(s) + CO_2(g) \rightleftharpoons 2CO(g) \quad K_p(2) = 1.3 \times 10^{14}$$

13. Consider the following equilibrium:

$$CO(g) + \tfrac{1}{2}O_2(g) \rightleftharpoons CO_2(g)$$

Calculate K_p for this equilibrium using the following equations.

$$C(s) + \tfrac{1}{2}O_2(g) \rightleftharpoons CO(g) \quad K_p = 2.9 \times 10^{10}$$

$$C(s) + O_2 \rightleftharpoons CO_2(g) \quad K_p = 4.8 \times 10^{20}$$

14. Consider the equilibrium in which methanol, CH_3OH, results from carbon monoxide, CO, and hydrogen, H_2:

$$CO(g) + 2H_2(g) \rightleftharpoons CH_3OH(g)$$

The equilibrium constant K_c for this reaction is 290 at 700 K. From what direction does the equilibrium establish, if the initial concentrations are 1.0 M CO, 1.0 M H_2, and 2.0 M CH_3OH?

15. The equilibrium constant K_p for the following reaction is 4.5×10^{-5} at 723 K.

$$N_2(g) + 3H_2(g) \rightleftharpoons 2NH_3(g)$$

Determine the direction equilibrium establishes when initially 20 atm of N_2, 15 atm of H_2, and 5.0 atm of NH_3 are present in the reaction vessel.

16. Consider the following equilibrium:

$$H_2(g) + I_2(g) \rightleftharpoons 2HI(g) \quad K_c = 45.9 \ (763 \text{ K})$$

If the initial concentrations of 0.100 M H_2 and 0.100 M I_2 are allowed to reach equilibrium, what are the equilibrium concentrations of all components of the equilibrium?

17. For the reaction

$$PCl_3(g) + Cl_2(g) \rightleftharpoons PCl_5(g)$$

the equilibrium constant K_c is 13.7 at 546 K. What are the equilibrium concentrations of all components of the equilibrium if 0.350 M PCl_3, 0.350 M Cl_2, and 0.100 M PCl_5 are present initially in the reaction vessel?

18. The K_p value for

$$2PH_3(g) \rightleftharpoons P_2(g) + 3H_2(g)$$

is 1.04×10^{-6} at 500 K. Initially 2.75 atm PH_3 is introduced into a reaction vessel and the system is allowed to reach equilibrium. What are the equilibrium partial pressures of all components of the system?

19. Dinitrogen tetroxide, N_2O_4, molecules dissociate to produce nitrogen dioxide, NO_2. The equilibrium constant K_c for

$$N_2O_4(g) \rightleftharpoons 2NO_2(g)$$

is 0.125 at 298 K. What percent of the N_2O_4 dissociates when a 5.09-g sample of N_2O_4 is added to a 3.19-L reaction vessel?

20. State Le Chatelier's principle.

For Problems 21 to 24, consider the following equilibrium:

$$2NO(g) + O_2(g) \rightleftharpoons 2NO_2(g) \quad \Delta H = -117 \text{ kJ}$$

21. Predict the effect of an increase in the concentration of NO on the equilibrium concentration of NO_2.

22. Predict the effect of a pressure decrease as a result of increased volume on the equilibrium concentration of NO_2.

23. Predict the effect of decreasing the volume on the equilibrium concentration of NO_2.

24. Predict the effect of decreasing the temperature on the equilibrium concentration of NO_2.

25. What equation is used to calculate the effect of a temperature change on the magnitude of the equilibrium constant?

26. a. Write the equilibrium expression, K_c, for the following:

    $$CaSO_3(s) \rightleftharpoons CaO(s) + O_2(g)$$

    b. How would the equilibrium constant be measured for this equilibrium?

27. The equilibrium constant K_c for the reaction

    $$2CO_2(g) \rightleftharpoons 2CO(g) + O_2(g)$$

    is 33.3 at 1030 K. What is the K_p value of the following?

    $$2CO(g) + O_2(g) \rightleftharpoons 2CO_2(g)$$

28. Consider the following equilibrium:

    $$PCl_5(g) \rightleftharpoons PCl_3(g) + Cl_2(g)$$

    Initially 0.0646 M PCl_5 is introduced into a reaction vessel 350 K. At equilibrium 0.0444 M Cl_2 is found. Calculate K_c.

29. The equilibrium constant K_p for

    $$2ClF_3(g) \rightleftharpoons Cl_2(g) + 3F_2(g)$$

    is 0.22 at 1100 K. If an equilibrium mixture contains 0.61 atm Cl_2 and 0.23 atm ClF_3, what is the equilibrium partial pressure of F_2?

30. Calculate the equilibrium constant for the following equilibrium at 2000 K:

    $$BrF_3(g) \rightleftharpoons BrF(g) + F_2(g)$$

    given the following equations and equilibrium constants.

    $$Br_2(g) + F_2(g) \rightleftharpoons 2BrF(g) \quad K_p = 2.2 \times 10^4$$

    $$\tfrac{1}{2}Br_2(g) + 3/2F_2(g) \rightleftharpoons BrF_3(g) \quad K_p = 2.3$$

31. The equilibrium constant K_c for

    $$CO_2(g) + H_2(g) \rightleftharpoons H_2O(g) + CO(g)$$

    is 302 at 600 K. For the following initial concentrations, determine if the system is in equilibrium, and if not, the direction the equilibrium develops.

    $$[CO_2] = [H_2] = 0.489\ M$$

    $$[H_2O] = [CO] = 2.71\ M$$

32. At 673 K the equilibrium constant K_c for

    $$Fe_2O_3(s) + 3H_2(g) \rightleftharpoons 2Fe(s) + 3H_2O(g)$$

    is 0.35. What are the equilibrium concentrations of H_2 and H_2O if 5.0 M H_2, 250.0 g Fe_2O_3, and 250.0 g Fe are initially added to the reaction vessel?

33. At elevated temperatures solid ammonium chloride, $NH_4Cl(s)$, dissociates to $NH_3(g)$ and $HCl(g)$.

    $$NH_4Cl(s) \rightleftharpoons NH_3(g) + HCl(g)$$

    When a sample of ammonium chloride is allowed to equilibrate at 613 K, the total pressure becomes 1.00 atm. What is the K_p for the reaction?

34. What is the difference between the K and Q expressions?

35. Consider the following equilibrium:

    $$N_2(g) + 3H_2(g) \rightleftharpoons 2NH_3(g) \quad K_c = 0.0060\ (298\ K)$$

    Determine the equilibrium constants, K_c for the following.

    $$NH_3(g) \rightleftharpoons \tfrac{1}{2}N_2(g) + 3/2H_2(g)$$

    a. 1.7×10^2

    b. 13

    c. 0.077

    d. 0.0030

    e. none of these

36. The following gas-phase equilibrium has a K_p value of 1.78 at 523 K:

    $$PCl_5(g) \rightleftharpoons PCl_3(g) + Cl_2(g)$$

    Calculate the value of K_c.

    a. 0.0415

b. 76.4

c. 4.09×10^{-4}

d. 0.0728

e. none of these

37. Which of the following statements is false?

    a. At equilibrium the value of Q equals K.

    b. Reactions with larger equilibrium constants occur rapidly.

    c. Most equilibria with small values of K have low concentrations of products at equilibrium

    d. If Q is greater than K, the equilibrium establishes from right to left.

For Problems 38 to 40, consider the following equilibrium:

$$4HCl(g) + O_2(g) \rightleftarrows 2H_2O(g) + 2Cl_2(g)$$

$$\Delta H = -116 \text{ kJ}$$

38. What effect does the addition of H_2O have on the equilibrium concentration of Cl_2?

    a. $[Cl_2]$ increases

    b. $[Cl_2]$ remains constant

    c. $[Cl_2]$ decreases

39. What effect does the removal of HCl have on the equilibrium concentration of Cl_2?

    a. $[Cl_2]$ increases

    b. $[Cl_2]$ remains constant

    c. $[Cl_2]$ decreases

40. What effect does the addition of a catalyst have on the equilibrium concentration of Cl_2?

    a. $[Cl_2]$ increases

    b. $[Cl_2]$ remains constant

    c. $[Cl_2]$ decreases

Check Yourself

1. $K_c = ([C]^y[D]^z)/([A]^w[B]^x)$ **(Equilibrium expressions)**

2. $K_c = ([CO][H_2O])/([H_2][CO_2])$

   $K_c = (0.0390\ M)(0.0390\ M)/((0.0609\ M)(0.0609\ M)) = 0.410$ **(Equilibrium expressions)**

3. $K_c = \dfrac{[CO]^2}{[CO_2]}$ **(Equilibrium expressions)**

4. $K_c = [N_2][H_2O]^3/[NH_3]^2$ **(Equilibrium expressions)**

5. a. $K_c = [NH_3]^2/([N_2][H_2]^3) = 0.51$

   b. $K_{reverse} = 1/K_c = ([N_2][H_2]^3)/[NH_3]^2 = 1/0.51 = 2.0$. **(Equilibrium constants)**

6. $K_c = (1/K_c)^2 = (1/20.4)^2 = 0.00240$ **(Equilibrium constants)**

7. $K_p = \dfrac{P^2_{NO_2}}{P_{N_2O_4}}$ **(Equilibrium expressions)**

8. $K_p = K_c(RT)^{\Delta n}$

   $\Delta n = 1 - 2 = -1$

   $K_p = 4.6 \times 10^9 \ (0.0821 \text{ L atm/mol K} \times 373 \text{ K})^{-1}$

   $= 4.6 \times 10^9 \times 1/(0.0821 \text{ L atm/mol K} \times 373 \text{ K}) = 1.5 \times 10^8$ **(K_p- K_c relationship)**

9. a. $[H_2]_{reacted} = [H_2]_{initial} - [H_2]_{final} = 0.250\ M - 0.0314\ M = 0.219\ M$

   $[HCl] = 2 \times 0.219\ M = 0.438\ M$

   $K_c = [HCl]^2/([H_2][Cl_2]) = (0.438\ M)^2/(0.0314\ M \times 0.0314\ M) = 195$

   b. $K_p = K_c(RT)^n$

Because the number of moles of gaseous products (2HCl) equals that of gaseous reactants ($H_2 + Cl_2$), the Δn for the equilibrium is 0. When the RT term is raised to the zero power, 1 is obtained, which means that K_p equals K_c. The K_p for the equilibrium is 195. **(Equilibrium expressions)**

10. $n = 3.35$ g $SO_2Cl_2 \times 1$ mol $SO_2Cl_2/135$ g $SO_2Cl_2 = 0.0248$ mol SO_2Cl_2

    $PV = nRT$

    $P_{SO_2Cl_2} = (nRT)/V = (0.0248$ mol $\times 0.0821$ L·atm/mol·K $\times 300$ K$)/1.00$ L $= 0.611$ atm

    $P_{Cl_2} = P_{SO_2} = 0.120$ atm

    $P_{SO_2Cl_2} = 0.611$ atm $- 0.120$ atm $= 0.491$ atm

    $K_p = (P_{SO_2} \cdot P_{Cl_2})/P_{SO_2Cl_2} = (0.120$ atm $\times 0.120$ atm$)/0.491$ atm $= 0.0293$ **(Equilibrium constants)**

11. $K_c = K_p/(RT)^{\Delta n} = 0.0293/(0.0821$ L·atm/mol·K $\times 300$ K$) = 1.19 \times 10^{-3}$ **(Equilibrium constants)**

12. First, multiply the first equation by 2 so that the CO cancel when added.

    $2CO(g) + 2Cl_2(g) \rightleftarrows 2COCl_2(g)$

    $+ \ C(s) + CO_2(g) \rightleftarrows 2CO(g)$

    $C(s) + CO_2(g) + 2Cl_2(g) \rightleftarrows 2COCl_2(g)$

    When the first equation is doubled, the equilibrium constant is changed to $(6.0 \times 10^{-3})^2$ or 3.6×10^{-5}, because each term in the equation now has a coefficient of 2, the equilibrium constant must be squared.

    $K_p = (3.6 \times 10^{-5}) \times (1.3 \times 10^{14}) = 4.7 \times 10^9$ **(Rule of multiple equilibria)**

13. $CO(g) \rightleftarrows C(s) + \frac{1}{2}O_2(g) \quad K_p = 1/2.9 \times 10^{10} = 3.4 \times 10^{-11}$

    $C(s) + O_2 \rightleftarrows CO_2(g) \quad K_p = 4.8 \times 10^{20}$

    $K_p = 3.4 \times 10^{-11} \times 4.8 \times 10^{20} = 1.6 \times 10^{10}$ **(Rule of multiple equilibria)**

14. $Q = [CH_3OH]/([CO][H_2]^2) = 2.0\ M/(1.0\ M \times 1.0\ M^2) = 2$

    Because the value of Q (2) is less than that of K (290), the reaction proceeds from left to right to reach equilibrium. **(Reaction quotient)**

15. $Q = P^2_{NH_3}/P_{N_2}P^3_{H_2} = (5.0$ atm$)^2/(20$ atm $\times (15$ atm$)^3) = 3.70 \times 10^{-4}$

    Because the value of Q, 3.70×10^{-4}, is larger than the value of K_p, 4.5×10^{-5}, the system approaches equilibrium from right to left. **(Reaction quotient)**

16.

	$H_2(g)$ +	$I_2(g) \rightleftarrows$	$2HI(g)$
Initial	0.100 M	0.100 M	0.0 M
Change	$-x$	$-x$	$+2x$
Equilibrium	0.100 $M - x$	0.100 $M - x$	$2x$

$K_c = [HI]^2/[H_2][I_2]$

$45.9 = (2x)^2/(0.100 - x)^2$

$\pm 6.77 = 2x/(0.100 - x)$

$x = 0.0772\ M$ and ($0.142\ M$–not possible because only $0.100\ M$ is initially present)

Using the first value, both [H$_2$] and [I$_2$] equal 0.100 – x, thus they are 0.100 M – 0.0772 M or 0.023 M. At equilibrium [HI] equals 2x or 2 × 0.0772 M, which is 0.154 M. **(Equilibrium problem)**

17.

Compound	Initial Concentration, M	Concentration Change	Equilibrium Concentration, M
[PCl$_3$]	0.350	–x	0.350 –x
[Cl$_2$]	0.350	–x	0.350 –x
[PCl$_5$]	0.100	+x	0.100 + x

K_c = [PCl$_5$]/([PCl$_3$][Cl$_2$])

13.7 = (0.100 + x)/(0.350 – x)^2

13.7 = (0.100 + x)/(0.123 – 0.700x + x^2)

13.7x^2 –9.59x + 1.68 = 0.100 + x

13.7x^2 –10.59x + 1.58 = 0

Using the quadratic equation gives 0.202 M and 0.571 M. Only 0.202 M has physical reality; thus,

[PCl$_3$] = 0.350 – x = 0.350 M – 0.202 M = 0.148 M

[Cl$_2$] = 0.350 – x = 0.350 M – 0.202 M = 0.148 M

[PCl$_5$] = 0.100 + x = 0.100 M + 0.202 M = 0.302 M   **(Equilibrium problem)**

18.

Compound	Initial Pressure, atm	Pressure Change	Equilibrium Pressure, atm
P_{PH_3}	2.75	–2x	2.75 – 2x
P_{P_2}	0	+x	x
P_{H_2}	0	+3x	3x

$K_p = P_{P_2} P_{H_2}^3 / P_{PH_3}^2$

1.04 × 10^{-6} = x (3x)^3/(2.75 –2x)^2 = 27x^4/(2.75 –2x)^2

x = 0.0232 atm

P_{PH_3} = 2.75 atm – 2x = 2.75 atm – 2(0.0230 atm) = 2.70 atm

P_{P_2} = x = 0.0230 atm

P_{H_2} = 3x = 3(0.0230 atm) = 0.0690 atm   **(Equilibrium problem)**

19. [N$_2$O$_4$] = (5.09 g N$_2$O$_4$ × 1 mol N$_2$O$_4$/92.0 g N$_2$O$_4$)/3.19 L = 0.0173 M

Compound	Initial Concentration, M	Concentration Change	Equilibrium Concentration, M
[N$_2$O$_4$]	0.0173	–x	0.0173 – x
[NO$_2$]	0	+2x	2x

$K_c = [NO_2]^2/[N_2O_4]$

$0.125 = (2x)^2/0.0173 - x$

$x = 0.0124\ M$

% dissociation = $(M(N_2O_4$ dissociated$)/M(N_2O_4$ undissociated$)) \times 100$

% dissociation = $(0.0124\ M/0.0173\ M) \times 100 = 71.7\%$ **(Equilibrium problem)**

20. Le Chatelier's principle states that equilibrium systems tend to shift in such a way as to reduce changes in temperature, pressure, and concentration. **(Le Chatelier's principle)**

21. When a reactant concentration is increased such as NO, it shifts the equilibrium to the right to absorb the increased number of NO molecules; therefore, the $[NO_2]$ increases. **(Le Chatelier's principle)**

22. A pressure decrease that results in an increase in volume favors the reaction that yields the larger number of molecules. Because the reverse reaction produces three moles and the forward reaction produces two moles, the reverse reaction is favored; accordingly, the $[NO_2]$ decreases. **(Le Chatelier's principle)**

23. The volume decreases when the pressure increases. At smaller volumes and higher pressures the reaction favors the side of the reaction that produces a smaller number of moles of particles. The forward reaction, which produces two moles of molecules, decreases the total number of moles in the system; therefore, the $[NO_2]$ increases. **(Le Chatelier's principle)**

24. The enthalpy change for the reaction is -117 kJ, which means the forward reaction is exothermic and the reverse reaction is endothermic. A temperature decrease favors the exothermic reaction because it replaces the heat removed; subsequently, the $[NO_2]$ increases. **(Le Chatelier's principle)**

25. $\ln(K_2/K_1) = (\Delta H/R)((T_2 - T_1)/(T_1 T_2))$ **(Le Chatelier's principle)**

26. a. $K_c = [O_2]$ b. Measure the molar concentration of O_2. **(Equilibrium expression)**

27. $K_p = K_c(RT)^{\Delta n} = 33.3 \times (0.0821\ (L \cdot atm)/(mol \cdot K) \times 1030\ K)^1 = 2.82 \times 10^3$ **(Equilibrium constant)**

28. $K_c = [PCl_3][Cl_2]/[PCl_5] = (0.0444\ M)^2/(0.646\ M - 0.0444\ M) = 0.0976$

    $K_p = K_c(RT)^n = 0.0976 \times ((0.0821\ (L \cdot atm)/(mol \cdot K) \times 350\ K)^1 = 2.80$ **(Equilibrium constant)**

29. $K_p = P_{Cl_2} P_{F_2}^3 / P_{Cl_2F_3}^2 = 0.61\ atm \times P_{F_2}^3/(0.23\ atm)^2 = 0.22$

    $P_{F_2} = 0.27$ atm **(Equilibrium constant)**

30. ½Br₂ + ½F₂ ⇌ BrF $K_p = (2.2 \times 10^4)^{0.5}$

    + BrF₃ ⇌ ½Br₂ + 3/2 F₂ $K_p = 1/2.3$

    BrF₃ ⇌ BrF + F₂ $K_p = (2.2 \times 10^4)^{0.5} \times 1/2.3 = 65$

    (Rule of multiple equilibria)

31. $K_c = ([CO][H_2O])/([H_2][CO_2]) = 302$

    $Q = (2.71\ M)^2/(0.489\ M)^2 = 30.7$

    $Q < K_c$, thus the equilibrium establishes from left to right. **(Reaction quotient)**

32.

Compound	Initial Concentration, M	Concentration Change	Equilibrium Concentration, M
$[H_2]$	5.0	$-3x$	$5.0 - 3x$
$[H_2O]$	0	$+3x$	$3x$

$K_c = [H_2O]^3/[H_2]^3 = 0.35$

$(3x)^3/(5.0-3x)^3 = 0.35$

$x = 0.69\ M$; thus, $[H_2] = 5.0 - 3x = 2.93\ M$ and $[H_2O] = 3x = 2.1\ M$ **(Equilibrium problem)**

33. $P_{NH_3} + P_{HCl} = x$

$P_{total} = P_{NH_3} + P_{HCl} = 1.00\ atm$

$2x = 1.00$

$x = 0.50\ atm$

$K_p = P_{NH_3}P_{HCl} = (0.50\ atm)^2 = 0.25$ **(Equilibrium constants)**

34. The Q and K expressions take the same form. The K expression shows the relationship of concentrations when the system is at equilibrium. The Q expression is used when the system is not at equilibrium. (*K* and *Q* expressions)

35. b. $K_c = 1/(0.0060)^{0.5} = 13$ **(Equilibrium constant)**

36. a. $K_p = K_c(RT)^{\Delta n}$

    $K_c = K_p/(RT)^n = 1.78/((0.0821\ (L\cdot atm)/(mol\cdot K) \times 523\ K)^1 = 0.0415$ (K_p-K_c relationship)

37. b. "Reactions with larger equilibrium constants occur rapidly" is false because the rate of a reaction is independent of the value of K. (*K* and *Q* expressions)

38. c. $[Cl_2]$ decreases because the addition of water shifts the equilibrium to the left. **(Le Chatelier's principle)**

39. c. $[Cl_2]$ decreases because the removal of water shifts the equilibrium to the left. **(Le Chatelier's principle)**

40. b. $[Cl_2]$ remains constant because a catalyst has no effect on the equilibrium concentrations. **(Le Chatelier's principle)**

Grade Yourself

Circle the numbers of the questions you missed, then fill in the total incorrect for each topic. If you answered more than three questions incorrectly, you need to focus on that topic. (If a topic has less than three questions and you had at least one wrong, we suggest you study that topic also. Read your textbook, a review book, or ask your teacher for help.)

Subject: Fundamental Principles of Chemical Equilibria

Topic	Question Numbers	Number Incorrect
Equilibrium expressions	1, 2, 3, 4, 7, 9, 26	
Equilibrium constants	5, 6, 10, 11, 27, 28, 29, 33, 35	
K_p - K_c relationship	8, 36	
Rule of multiple equilibria	12, 13, 30	
Reaction quotient	14, 15, 31	
Equilibrium problem	16, 17, 18, 19, 32	
Le Chatelier's principle	20, 21, 22, 23, 24, 25, 38, 39, 40	
K and Q expressions	34, 37	

Acids, Bases, and Salts

15

Brief Yourself

Acids are substances that taste sour, react with many metals to liberate H_2, change the color of acid-base indicators opposite to that of bases, and neutralize bases. Bases are substances that taste bitter, have a slippery feeling, change the color of acid-base indicators opposite to that of acids, and neutralize acids.

Arrhenius defined acids as substances that increase the hydrogen ion concentration in aqueous solution, and bases as substances that increase the hydroxide ion concentration in aqueous solution. When a hydrogen ion enters aqueous solution it combines with a water molecule and produces a hydronium ion, H_3O^+, which is hydrated by additional water molecules. It is most common to represent dissolved hydrogen ions as $H^+(aq)$. When an Arrhenius acid combines with an Arrhenius base, a neutralization reaction occurs.

The Brønsted-Lowry definitions define acids as proton donors and bases as proton acceptors. When an acid, HA, releases a proton it produces its conjugate base, A^-

$$HA \rightarrow H^+ + A^-$$

Acid   Conjugate Base

After a conjugate base accepts a proton it re-forms the acid. When a base, B, accepts a proton it produces its conjugate acid, HB^+.

$$B + H^+ \rightarrow HB^+$$

Base   Conjugate Acid

After a conjugate acid releases a proton it re-forms the base. The conjugate bases of weak acids tend to be relatively strong bases and the conjugate acids of weak bases tend to be relatively strong acids. A chemical species is amphiprotic if it either donates or accepts protons, and a chemical species is amphoteric if it reacts with both acids and bases.

Strong acids almost totally ionize in dilute solutions and weak acids only partially ionize, usually less than 10%. Proton-donating tendency is related to the strength of the conjugate base. Stronger acids have weaker conjugate bases than weaker acids. It is generally true that the strongest Brønsted-Lowry acid that exists in a solution is the conjugate acid of the solvent. Therefore the strongest acid in aqueous solution is the hydronium ion, H_3O^+. The effect that acids with different tendencies to donate protons produce solutions with the same acidic properties is called the leveling effect. The strengths of bases are measured in terms of their proton-accepting tendency. A stronger base has a greater tendency to accept protons than weaker bases. Acid-base strength can also be predicted from the structure of molecules. Molecules that have the capacity to withdraw electron density from the bond to the ionizable H atom tend to be acidic. Factors that stabilize the resulting anion also help to increase acid strength.

A Lewis acid is an electron-pair acceptor and a Lewis base is an electron pair donor. After a Lewis acid accepts an electron pair from a Lewis base it forms an acid-base adduct. Lewis acids are electron deficient species such as cations, molecules that have an atom with less than eight outer-level electrons, and molecules that can expand their octets. Lewis bases are electron-rich species such as anions and molecules with atoms with one or more lone pairs.

Some acids are prepared by mixing nonmetal oxides, also called acid anhydrides, with water, and others by direct combination of nonmetals with hydrogen. Bases result when metal oxides, also called basic anhydrides, react with water. Sulfuric acid, hydrochloric acid, nitric acid, phosphoric acid, and perchloric acid are five of the most important industrial acids. Sodium hydroxide, potassium hydroxide, and ammonia are three of the most significant industrial bases.

When equivalent amounts react, strong acids neutralize strong bases and produce neutral solutions. Equivalent amounts of strong bases and weak acids react, producing basic solutions. Equivalent amounts of weak bases neutralize strong acids to produce acidic solutions. Weak bases and weak acids combine to produce neutral, acidic, or basic solutions, depending on the relative strength of the reactants.

Salts are ionic compounds that result when acids combine with bases. Each salt is composed of a cation other than H^+ and an anion other than O^{2-} or OH^-. In a dilute aqueous solution, soluble salts almost totally dissociate into ions. Salts may be acidic, neutral, or basic. The acid-base properties of salts are predicted from the type of neutralization reaction that occurs to produce the salts. Neutral salts result from the reaction of strong acids and strong bases, basic salts result from the reaction of strong bases and weak acids, and acidic salts result from the reaction of strong acids and weak bases.

Test Yourself

1. List the principal properties of acids.

2. Define Arrhenius acids and bases.

3. Write an equation to show that ammonia, NH_3, is an Arrhenius base.

4. a. How is a hydrogen ion different from a proton? b. Explain the nature of hydrogen ions in aqueous solution.

5. a. What are Brønsted-Lowry acids and bases? b. Write equations that show the behavior of Brønsted-Lowry acids and bases.

6. Write three equations that show the proton donating ability of the triprotic acid, phosphoric acid, H_3PO_4.

7. What are the conjugate bases of HNO_3, NH_3, and OH^-?

8. What are the conjugate acids of HNO_3, NH_3, OH^-?

9. Write the equation for the reaction of hydrofluoric acid and ammonia and identify the acid, base, conjugate acid, and conjugate base.

10. Write two equations that show the amphiprotic property of water.

11. a. What is the difference between an amphiprotic and amphoteric substance? b. Give an example of an amphoteric substance that is not amphiprotic.

12. In which direction, right or left, does the following equilibrium lie? Explain.

$$HNO_2 + CN^- \rightleftarrows NO_2^- + HCN$$

13. Arrange the following acids from strongest to weakest: CH_3COOH, $CClH_2COOH$, CCl_2HCOOH, and CCl_3COOH. Explain.

14. a. What are Lewis acids and bases? b. What are the general characteristics of Lewis acids and bases?

15. Show that the reaction of Ag^+ and Cl^- is a Lewis acid-base reaction. Identify the acid and base.

16. Consider the following reaction:

$$Cd^{2+}(aq) + 4I^-(aq) \rightleftarrows CdI_4^{2-}$$

Which reactant is the Lewis acid and base. Explain.

17. Consider the following equilibria:

$$NH_3 + H_2O \rightleftarrows NH_4^+ + OH^- \quad K = 1.8 \times 10^{-5}$$

$$CH_3NH_2 + H_2O \rightleftarrows CH_3NH_3+ + OH^-$$
$$K = 4.4 \times 10^{-4}$$

Which is a stronger base, NH_3 or CH_3NH_2? Explain.

18. Write an equation that shows how phosphoric acid could be synthesized.

19. a. What is a neutralization reaction? b. Write an example of an equation for a neutralization reaction.

20. a. Write the overall equation for the neutralization of HBr by NaOH. b. Write the overall ionic equation for this reaction. c. What is the net ionic equation?

21. What is the net ionic equation for the reaction of acetic acid, CH_3COOH, and potassium hydroxide?

22. Write the overall and net ionic equations for the neutralization of aqueous ammonia with hydrochloric acid.

23. Write the overall ionic and net ionic equation for the reaction of aqueous rubidium hydroxide and nitric acid. b. When equivalent amounts are mixed, is the solution acidic, basic, or neutral? Explain.

24. What are salts? How is a salt produced?

25. a. What salt results when aqueous ammonia and hydrochloric acid react? b. Write the equation for this reaction? c. Is this salt neutral, acidic, or basic?

26. Is the salt potassium cyanide an acidic, basic, or neutral salt? Write an equation to support your answer.

27. When the oxide ion, O^{2-}, reacts with water it produces hydroxide ions. a. Write the equation for this reaction. b. What are the acid and base in this reaction? c. What are the conjugate acid and conjugate base in this reaction?

28. Write two equations that show the amphiprotic behavior of the hydrogencarbonate ion, HCO_3^-.

29. What is the leveling effect of acids and bases in water?

30. Arrange the following compounds in order of increasing acid strength: H_2Se, H_2Te, H_2S, and H_2O. Explain.

31. Ammonia, NH_3, exhibits basic properties in water but nitrogen trifluoride, NF_3, is not basic.

Write an explanation for this difference in basic properties.

32. Aluminum ions, Al^{3+}, react with water to produce hydronium and $[Al(H_2O)_5OH]^{2+}$ ions; thus, solutions containing Al^{3+} are acidic. Sodium ions, Na^+, are only hydrated when they enter aqueous solution; hence, they form neutral solutions. What may account for the difference in their behavior?

33. In terms of their structures, explain why methoxide ions, CH_3O^-, are stronger bases than hydroxide ions, OH^-.

34. Solid nickel reacts with carbon monoxide and produces nickel tetracarbonyl, $Ni(CO)_4$. a. Write an equation for the reaction. b. Identify the Lewis acid and base in this reaction.

35. Which of the following is the strongest acid?
    a. CH_4
    b. NH_3
    c. H_2O
    d. HF

36. Which of the following is the weakest acid?
    a. $HClO_4$
    b. $HClO_3$
    c. $HClO_2$
    d. HClO
    e. none of these

37. Which of the following shows the acids listed from strongest to weakest?
    a. HOCl > HOBr > HOI
    b. HOBr > HOCl > HOI
    c. HOI > HOBr > HOCl
    d. HOI > HOCl > HOBr
    e. none of these

38. Which of the following is the strongest Lewis acid?
    a. CCl_4
    b. $AlCl_3$
    c. NCl_3
    d. OCl_2
    e. none of these are Lewis acids

39. Which of the following acids are classified as Lewis acids but not are not Brønsted-Lowry acids?
    a. $HBrO_3$
    b. $SbCl_3$
    c. HSO_4^-
    d. AlF_3
    e. none of these

40. Which of the following acids is one of the most important commercial chemicals and is a strong dehydrating agent?
    a. $HCl(aq)$
    b. $HNO_3(aq)$
    c. $H_2SO_4(aq)$
    d. $HClO_4(aq)$
    e. none of these

Check Yourself

1. Acids taste sour. They change the color of acid-base indicators differently from that of bases. For example, acids change litmus from blue to red, while bases change litmus from red to blue. Acids and bases neutralize each other and produce salts. Many acids react with metals and release hydrogen gas. (**Acid-base properties**)

2. Arrhenius acids increase the hydrogen ion (H^+) concentration in water. An Arrhenius base increases the hydroxide ion (OH^-) concentration in water. (**Arrhenius acids**)

3. $NH_3(aq) + H_2O(l) \rightleftarrows NH_4^+(aq) + OH^-(aq)$ (**Arrhenius acids**)

4. A hydrogen ion is just a proton–a very small body with a high charge density. It does not exists by itself in the hydrogen bonded network of water molecules. An aqueous hydrogen ion is that it combines with a water molecule to produce a hydronium ion, H_3O^+. (**Nature of acids**)

$$H^+(aq) + H_2O(l) \rightarrow H_3O^+(aq)$$

5. a. A Brønsted-Lowry acid is a proton donor and a base is a proton acceptor.

   b. $HCl(g) + NH_3(g) \rightarrow NH_4Cl(s)$
       acid   base

   HCl donates a proton (acid) and NH_3 accepts a proton (base). (**Brønsted-Lowry acids-bases**)

6. $H_3PO_4(aq) \rightarrow H^+(aq) + H_2PO_4^-(aq)$

   $H_2PO_4^-(aq) \rightarrow H^+(aq) + HPO_4^{2-}(aq)$

   $H_2PO_4^{2-}(aq) \rightarrow H^+(aq) + PO_4^{3-}(aq)$ (**Polyprotic acids**)

7.

Acid	Conjugate Base
HNO_3	NO_3^-
NH_3	NH_2^-
OH^-	O^{2-}

   (**Brønsted-Lowry acids-bases**)

8.

Base	Conjugate Acid
HNO_3	$H_2NO_3^+$
NH_3	NH_4^+
OH^-	H_2O

   (**Brønsted-Lowry acids-bases**)

9. $HF + NH_3 \rightleftarrows F^- + NH_4^+$
   acid base   conjugate base conjugate acid (**Brønsted-Lowry acids-bases**)

10. $H_2O(l) + NH_3(aq) \rightleftarrows OH^-(aq) + NH_4^+(aq)$
    acid

    $H_2O(l) + HCl(g) \rightarrow H_3O^+(aq) + Cl^-(aq)$ (**Brønsted-Lowry acids-bases**)
    base

11. a. An amphoteric species is one that can act as either an acid or a base. An amphiprotic substance can donate or accept protons. b. Some substances are acids and bases but do not donate or accept protons; e.g., consider aluminum oxide.

    $Al_2O_3(s) + 6HCl(aq) \rightarrow 2AlCl_3(aq) + 3H_2O(l)$
    base

    $Al_2O_3(s) + 2NaOH(aq) + 3H_2O(l) \rightarrow 2NaAl(OH)_4(aq)$
    acid   (**Brønsted-Lowry acids-bases**)

12. To predict the direction in which this equilibrium lies, consider the strengths of the acid and conjugate acid and the strengths of the base and conjugate base. HNO_2 is a stronger acid than HCN and CN^- is a stronger base than NO_2^-. The equilibrium favors the side with the weaker acid and base; hence, this equilibrium lies towards the right. (**Acid-base strength**)

13. $CCl_3COOH > CCl_2HCOOH > CClH_2COOH > CH_3COOH$

    As the number of Cl atoms increases more electron density is withdrawn from the O–H bond making it more polar which makes in more readily ionize in solution. In addition, the greater withdrawing capacity of Cl atoms helps stabilize the resulting negative charge in the conjugate base. (**Acid-base strength**)

14. a. A Lewis acid is an electron-pair acceptor. A Lewis base is an electron pair donor.

    b. Lewis acids are usually electron deficient. Lewis bases are electron rich. **(Lewis acids and bases)**

15. $Ag^+(aq) + Cl^-(aq) \rightarrow AgCl(s)$

    The Ag^+ accepts an electron pair from the Cl^-; thus, Ag^+ is a Lewis acid and Cl^- is a Lewis base. **(Lewis acids and bases)**

16. Lewis acid = Cd^{2+}, Lewis base = I^- Cd^{2+} accepts a pair of electrons from I^- in the formation of CdI_4^{2-}. **(Lewis acids and bases)**

17. The methylamine equilibrium lies farther to the right (because of its larger K value), indicating that it is a stronger base than ammonia. The methyl group in methylamine donates electron density into the positive ion, CH_3NH_3+. Thus, the decrease in charge stabilizes the ion and shifts the equilibrium farther to the right. **(Acid-base strength)**

18. $P_4O_{10}(s) + 6H_2O(l) \rightarrow 4H_3PO_4(aq)$ **(Preparation of acids and bases)**

19. a. A neutralization reaction occurs when equivalent amounts of an acid and base react with eachother.

    b. $HCl(aq) + NaOH(aq) \rightarrow NaCl(aq) + H_2O(l)$ **(Neutralization reactions)**

20. a. $HBr(aq) + NaOH(aq) \rightarrow NaBr(aq) + H_2O(l)$

    b. $H^+(aq) + Br^-(aq) + Na^+(aq) + OH^-(aq) \rightarrow Na^+(aq) + Br^-(aq) + H_2O(l)$

    c. $H^+(aq) + OH^-(aq) \rightarrow H_2O(l)$ **(Neutralization reactions)**

21. $CH_3COOH(aq) + OH^-(aq) \rightarrow CH_3COO^-(aq) + H_2O(l)$ **(Neutralization reactions)**

22. $H^+(aq) + Cl^-(aq) + NH_3(aq) \rightarrow NH_4^+(aq) + Cl^-(aq)$

    $H^+(aq) + NH_3(aq) \rightarrow NH_4^+(aq)$ **(Neutralization reactions)**

23. a. $H^+(aq) + NO_3^-(aq) + Rb^+(aq) + OH^-(aq) \rightarrow Rb^+(aq) + NO_3^-(aq) + H_2O(l)$

    $H^+(aq) + OH^-(aq) \rightarrow H_2O(l)$

    b. A neutral solution results because HNO_3 is a strong acid and RbOH is a strong base. **(Neutralization reactions)**

24. Salts are ionic compounds that result when acids combine with bases. Salts consist of a cation other than H^+ and an anion other than O^{2-} or OH^-. **(Salts)**

25. a. NH_4Cl, b. $NH_3(aq) + H^+(aq) + Cl^-(aq) \rightleftharpoons NH_4^+(aq) + Cl^-(aq)$, c. It is an acidic salt because the ammonium ion is a stronger acid than water–it is the conjugate acid of a weak base. **(Salts)**

26. Potassium ions are the result of the dissociation of the strong base KOH. Thus, they have little tendency to combine with water to produce an acidic solution–they are neutral cations. A cyanide ion is the conjugate base of the weak acid HCN. Thus, cyanide ions are relatively strong bases. A neutral cation and basic anion give a basic salt. The hydrolysis of water (splitting of water) by CN^- produces the hydroxide ions in solution.

    $$CN^-(aq) + H_2O(l) \rightleftharpoons HCN(aq) + OH^-(aq) \quad \textbf{(Salts)}$$

27. a. $O^{2-} + H_2O \rightarrow 2OH^-$ b. The acid is water, a proton donor, and the base is the oxide ion, a proton acceptor. c. Hydroxide is both the conjugate acid and conjugate base because both the water and oxide produce hydroxide in this reaction. **(Brønsted-Lowry acids-bases)**

28. $HCO_3^- + H^+ \rightarrow H_2CO_3$ **(Basic property)**

    $HCO_3^- + OH^- \rightarrow CO_3{2-}$ **(Acidic property)**

126 / College Chemistry

29. In water the strongest acid is the H⁺ and the strongest base is OH⁻. Thus, strong acids cannot be stronger than H⁺, and bases cannot be stronger than OH⁻. In other words, acids are leveled to H⁺, and bases to OH⁻. **(Acid-base strength)**

30. $H_2O > H_2S > H_2Se > H_2Te$ Oxygen is the most electronegative group 16 (VIA) element and forms the most polar bond with H; thus, it is the strongest acid. Tellurium is the least electronegative and forms the least polar bond with H; thus, it is the weakest acid. **(Acid-base strength)**

31. Ammonia, NH_3, has a lone pair that it can donate to Lewis acids. Adding three highly electronegative F atoms to the N diminishes the ability of N to donate the lone pair, making it much less basic. **(Acid-base strength)**

32. Al^{3+} ions have a much higher charge than Na^+ ions; therefore, they are significantly more acidic and can produce H_3O^+ and $[Al(H_2O)_5OH]^{2+}$. **(Lewis acid and bases)**

33. CH_3O^- ions are stronger bases than hydroxide ions because the methyl group, CH_3^-, can donate more electron density into the O atom, than can the H in OH⁻. The greater negative character in CH_3O^- makes it a better electron pair donor and proton acceptor. **(Acid-base strength)**

34. a. $4CO + Ni \rightarrow Ni(CO)_4$

    b. Ni, a metal, accepts one of the lone pair electrons from CO; hence, the Ni is a Lewis acid and CO is a Lewis base. **(Lewis acids and bases)**

35. d. HF is the strongest acid because it has the most polar bond which most readily ionizes in aqueous solution. **(Acid-base strength)**

36. d. HClO is the weakest acid because as the number of O atoms increases the O–H bond becomes more polar and more readily ionizes in aqueous solution. **(Acid-base strength)**

37. a. HOCl > HOBr > HOI

    The higher the electronegativity, the greater the electron density withdrawn from the O–H bond; therefore, Cl has the highest electronegativity and produces the strongest acid. **(Acid-base strength)**

38. b. $AlCl_3$ is the strongest Lewis acid because it is the only one that can readily accept an electron pair as a result of being electron deficient. **(Lewis acids and bases)**

39. b. and d. In this group, both $HBrO_3$ and HSO_4^- are Brønsted-Lowry acids because they can donate protons. $SbCl_3$ and AlF_3 are not Brønsted-Lowry acids because they cannot donate protons, but are Lewis acids because the central atoms can accept electron pairs. **(Lewis acids and bases)**

40. c. $H_2SO_4(aq)$ is the industrial chemical produced in greatest amount in the U.S. **(Important acids and bases)**

Grade Yourself

Circle the numbers of the questions you missed, then fill in the total incorrect for each topic. If you answered more than three questions incorrectly, you need to focus on that topic. (If a topic has less than three questions and you had at least one wrong, we suggest you study that topic also. Read your textbook, a review book, or ask your teacher for help.)

Subject: Acids, Bases, and Salts

Topic	Question Numbers	Number Incorrect
Acid-base properties	1	
Arrhenius acids	2, 3	
Brønsted-Lowry acids-bases	5, 7, 8, 9, 10, 11, 27	
Polyprotic acids	6	
Acid-base strength	12, 13, 17, 29, 30, 31, 33, 35, 36, 37	
Lewis acids and bases	14, 15, 16, 32, 34, 38, 39	
Preparation of acids and bases	18	
Neutralization reactions	19, 20, 21, 22, 23	
Salts	24, 25, 26	
Basic property	28a	
Acidic property	28b	
Important acids and bases Nature of acids	4, 40	

Acid-Base Equilibrium in Aqueous Solution

Brief Yourself

Water is a weak electrolyte and undergoes self-ionization to produce H^+ and OH^- ions. The ion-product equilibrium expression for water is

$$K_w = [H^+][OH^-] = 1.0 \times 10^{-14}$$

This equation is used to show the degree to which water ionizes. Because water is neutral, the molar concentrations of both the H^+ and OH^- ions are 1.0×10^{-7} M in pure water. Acids increase the H^+ ion concentration in water thus, according to Le Chatelier's principle, the OH^- ion concentration decreases. In all acidic solutions the H^+ ion concentration is greater than 10^{-7} M and the OH^- ion concentration is less than 10^{-7} M. In all basic solutions the H^+ ion concentration is less than 10^{-7} M and the OH^- ion concentration is greater than 10^{-7} M.

The pH scale is used to express the acidity or basicity of a solution. The pH of a solution is the negative common logarithm of the H^+ ion concentration.

$$pH = -\log [H^+]$$

Neutral solutions and pure water have a pH of 7, acidic solutions have pH values less than 7, and basic solutions have pH values above 7. Besides pH, the complementary pOH scale is also used. The pOH of a solution is the negative common logarithm of the OH^- ion concentration. Basic solutions have pOH values less than 7, and acidic solutions have values above 7. The sum of the pH and pOH for a given solution equals 14, the pK_w ($-\log K_w = 14$) for water. Acid-base indicators and pH meters are used to measure the pH of solutions.

Strong acids ionize 100% and thus for each mole of a monoprotic acid, HA, that dissolves, one mole of H^+ ions results. However, weak acids only ionize to a small degree, producing far fewer H^+ ions. Thus the solutions of weak acids have lower H^+ ion concentrations than strong acids of equal concentration. The degree to which weak acids ionize is expressed by writing the acid-ionization equilibrium expression. This expression gives us the acid-ionization equilibrium constant, K_a. A small value for K_a indicates a weak acid. The K_a values for strong acids are very large because strong acids totally ionize in dilute solutions. Acids are classified according to the number of number of H^+ ions they release per molecule. Monoprotic acids donate one proton, diprotic acids donate two protons, and triprotic acids donate three protons. Collectively, diprotic and triprotic acids are termed the polyprotic acids. After a polyprotic acid donates a proton, the resulting conjugate base has a proton that it can donate. Because the conjugate base has a negative charge, the K_a value for the second ionization is always much smaller.

Strong bases are substances that completely dissociate in solution and release OH^- ions. Examples of strong bases include NaOH and KOH. Most common weak bases produce OH^- ions by undergoing hydrolysis. For example, NH_3 undergoes hydrolysis in water and produces NH_4^+ and OH^- ions. The base-ionization equilibrium expression is used to show the degree to which NH_3 undergoes this reaction. This expression gives the base-ionization equilibrium constant, K_b. The product of the K_a of a weak acid and the K_b of its conjugate base equals the K_w of water. Thus either K_a or K_b can be calculated by dividing the other into K_w, 1.0×10^{-14}.

Salts are produced when acids undergo the neutralization reaction with bases. If a strong acid and base combine, a neutral salt results. If a strong acid and weak base react, an acidic salt is produced, and if a weak acid and strong base react, a basic salt results.

Test Yourself

1. a. Write the equation for the self-ionization of water. b. Write the equilibrium expression for this ionization.

2. a. What are the hydrogen and hydroxide ion concentration ranges in acidic solutions? b. What are the hydrogen and hydroxide ion concentration ranges in basic solutions? c. What are the hydrogen and hydroxide ion concentration in a neutral solution?

3. What is the hydroxide ion concentration in a solution that has $0.10\ M\ H^+$?

4. A 0.053-g sample of $Ca(OH)_2$ in dissolved in enough water to have 115 mL of solution. Calculate the $[H^+]$ and $[OH^-]$ for this solution.

5. Calculate the $[H^+]$ and $[OH^-]$ in a solution that contains 0.100 g HCl dissolved in 2.15 L of solution.

6. a. What is pH? b. What is the pH of a neutral solution?

7. What is the pH of a $0.1\ M\ OH^-$ solution?

8. A solution is prepared by dissolving 150 mg perchloric acid, $HClO_4$, in 3.55 L of solution. What is the pH of solution?

9. The pH of the blood is usually close to 7.35. What are the H^+ and OH^- ion concentrations in blood?

10. a. What is pOH? b. How is pOH related to pH?

11. What is the pH and pOH of a NaOH solution that has 0.055 g NaOH dissolved in 35 L solution?

12. Write the K_a expression for the following acid-ionization equilibrium.

$$HF(aq) \rightleftarrows H^+(aq) + F^-(aq)$$

13. A $0.010\ M$ butanoic acid solution is 3.9% ionized at 298 K. Calculate the K_a value for butanoic acid.

14. The pH of a $0.25\ M$ hypochlorous acid solution, $HOCl(aq)$, is 4.03. Calculate the K_a value for hypochlorous acid.

15. What is the percent ionization of a 1.0 M $HC_2H_3O_2$ solution? The K_a for $HC_2H_3O_2$ is 1.8×10^{-5}.

16. Calculate the equilibrium molar concentrations of lactic acid ($HC_3H_5O_3(aq)$), lactate, H^+, and pH of a 0.0010 M $HC_3H_5O_3$ solution. The K_a value for lactic acid is 1.4×10^{-4}.

$$HC_3H_5O_3(aq) \rightleftarrows H^+(aq) + C_3H_5O_3^-(aq)$$

17. Chloroacetic acid, $HC_2H_2O_2Cl$, is one of the stronger weak acids and has a K_a value of 1.4×10^{-3}. Calculate the pH of a 0.30 M solution.

18. Calculate the molar concentration of HF in a mixture of 0.100 M HF and 0.010 M HCl. The K_a value of HF is 7.4×10^{-4}.

19. Write the equations that show the ionizations of the diprotic acid sulfuric acid, H_2SO_4.

For Problems 20 to 24 consider the triprotic phosphoric acid, H_3PO_4. The acid-ionization constants for phosphoric acid are $K_{a1} = 7.5 \times 10^{-3}$, $K_{a2} = 6.2 \times 10^{-8}$, and $K_{a3} = 4.8 \times 10^{-13}$.

20. Write the equations for the ionizations of phosphoric acid.

21. What is the pH of 1.5 M H_3PO_4 solution?

22. What is the concentration of $H_2PO_4^-$ in this solution?

23. What is the concentration of HPO_4^{2-} in this solution?

24. What is the concentration of PO_4^{3-} in this solution?

25. a. Write the equation that shows the basic property of ammonia, NH_3. b. Write the base ionization equilibrium expression for ammonia, NH_3.

26. Calculate the pH and pOH of a 0.12 M hydroxylamine, H_2NOH, solution. The K_b for hydroxylamine is 1.1×10^{-8}.

27. The hypoiodite ion, IO^-, is the conjugate base of the acid hypoiodous acid, HIO. If the pH of a 0.030 M IO^- solution is 11.56, what is the K_a of hypoiodous acid.

28. Show that the sum of pK_a and pK_b equals 14.

29. If the pK_a of hypobromous acid, HBrO(aq) is 8.60, what is K_b of its conjugate base, BrO^-?

30. Calculate the pH of a 0.40 M calcium lactate, $Ca(C_3H_5O_3)_2$, solution. The K_a value for lactic acid, $HC_3H_5O_3$, is 1.4×10^{-4}.

31. A 0.19 M trifluoroacetic acid, $HCF_3CO_2(aq)$, solution is 79.6% ionized. Calculate the K_a value for trifluoroacetic acid.

32. Calculate the molarity of a benzoic acid, $HC_6H_5COO(aq)$, solution that has a pH of 2.89. The K_a of benzoic acid is 6.5×10^{-5}.

33. A solution of pure liquid ammonia, $NH_3(l)$, self ionizes. Write an equation that shows the self ionization of liquid ammonia.

34. What is the hydrogen ion concentration of a solution in which 0.015 mol HF is dissolved in 1.00 L of a 0.100 M NaF solution? The K_a of HF is 7.2×10^{-4}.

35. What is the pH of a solution that contains 0.10 mole H^+ dissolved in 10 L of solution?

a. 0.10

b. 1.00

c. 2.00

d. 10

e. none of these

36. Calculate the molar concentration of H^+ in 0.100 M HF. The K_a for HF is 7.2×10^{-4}.

a. 0.099 M

b. 2.10

c. 6.4×10^{-3} M

d. 0.0081 M

e. none of these

37. What is the hydrogen ion concentration in a solution that contains 4.0 g NaOH dissolved in 20.0 L solution?

    a. 0.0050 $M\ H^+$

    b. $2.0 \times 10^{-12}\ M\ H^+$

    c. $1.0 \times 10^{-14}\ M\ H^+$

    d. $5.0 \times 10^{-14}\ M\ H^+$

    e. none of these

38. Calculate the pH of a 0.75 M pyridine, C_5H_5N, solution. The K_b value for pyridine is 1.4×10^{-9}.

    a. 4.49

    b. 9.51

    c. 8.85

    d. 10.49

    e. none of these

39. If the K_a value of acetic acid, 1.8×10^{-5}, what is the K_b value of the acetate ion?

    a. 1.0×10^{-14}

    b. 1.8×10^{-5}

    c. 1.8×10^{-19}

    d. 5.6×10^{-10}

    e. cannot determine from this information

40. What is the pH of a solution that has a [OH$^-$] of $1.0 \times 10^{-10}\ M$?

    a. 10

    b. –10

    c. 1.0×10^{-4}

    d. 4

    e. none of these

Check Yourself

1. a. $H_2O(l) + H_2O(l) \rightleftarrows H_3O^+(aq) + OH^-(aq)$,

   b. $K_w = [H_3O^+][OH^-]$ or $K_w = [H^+][OH^-]$ **(Water self-ionization)**

2. a. $[H^+] > 1.0 \times 10^{-7}\ M$ and $[OH^-] < 1.0 \times 10^{-7}\ M$,

   b. $[H^+] < 1.0 \times 10^{-7}\ M$ and $[OH^-] > 1.0 \times 10^{-7}\ M$,

   c. $[H^+] = [OH^-] = 1.0 \times 10^{-7}\ M$ **(Water self-ionization)**

3. $1.0 \times 10^{-14} = [H^+][OH^-] = (0.10\ M\ H^+)[OH^-]$

   $[OH^-] = 1.0 \times 10^{-14}/0.10\ M\ H^+ = 1.0 \times 10^{-13}\ M$ **(H^+ and OH^- concentrations)**

4. $Ca(OH)_2(s) \xrightarrow{H_2O} Ca^{2+}(aq) + 2OH^-(aq)$

   0.053 g Ca(OH)$_2$ × (1 mol Ca(OH)$_2$/74.1 g Ca(OH)$_2$) × (2 mol OH$^-$/1 mol Ca(OH)$_2$) = 0.0014 mol OH$^-$

   [OH$^-$] = 0.0014 mol OH$^-$/(115 mL × 1 L/1000 mL) = 0.012 M

   $K_w = [H^+][OH^-]$

   $1.0 \times 10^{-14} = [H^+](0.012\ M\ OH^-)$

   $[H^+] = 1.0 \times 10^{-14}/0.012\ M = 8.3 \times 10^{-13}\ M\ H^+$ **(H^+ and OH^- concentrations)**

5. 0.100 g HCl × (1 mol HCl/36.5 g) × (1 mol H$^+$/1 mol HCl) = 0.0274 mol H$^+$

   $[H^+] = 0.0274$ mol/2.15 L = 0.00127 $M\ H^+$

   $1.0 \times 10^{-14} = (0.00127\ M\ H^+)[OH^-]$

   [OH$^-$] = $7.85 \times 10^{-12}\ M\ OH^-$ **(H^+ and OH^- concentrations)**

6. a. pH = –log [H$^+$]

   b. pH = –log [H$^+$] = –log ($1 \times 10^{-7}\ M\ H^+$) = 7.0 **(pH)**

7. $K_w = [H^+][OH^-]$

   $1.0 \times 10^{-14} = [H^+](0.1 \, M \, OH^-)$

   $[H^+] = 1.0 \times 10^{-14}/0.1 \, M = 1 \times 10^{-13} \, M$

   $pH = -\log[H^+] = -\log(1 \times 10^{-13} \, M) = 13.0$ **(pH)**

8. $150 \text{ mg HClO}_4 \times (1 \text{ g HClO}_4/1000 \text{ mg HClO}_4) \times (1 \text{ mol HClO}_4/100.5 \text{ g HClO}_4) \times$
   $(1 \text{ mol H}^+/1 \text{ mol HClO}_4) = 1.5 \times 10^{-3} \text{ mol H}^+$

   $[H^+] = 1.5 \times 10^{-3} \text{ mol H}^+/3.55 \, L = 4.2 \times 10^{-4} \, M$

   $pH = -\log[H^+] = -\log(4.2 \times 10^{-4} \, M) = 3.38$ **(pH)**

9. $pH = -\log[H^+]$

   $7.35 = -\log[H^+]$

   $-7.35 = \log[H^+]$

   $\text{antilog}(-7.35) = \text{antilog}(\log[H^+])$

   $4.5 \times 10^{-8} \, M = [H^+]$

   $[H^+] = 4.5 \times 10^{-8} \, M$

   $K_w = [H^+][OH^-]$

   $[OH^-] = K_w/[H^+] = 1.0 \times 10^{-14}/4.5 \times 10^{-8} \, M = 2.2 \times 10^{-7} \, M$ **(pH)**

10. a. $pOH = -\log[OH^-]$, b. $pK_w = 14 = pH + pOH$ **(pOH)**

11. $0.055 \text{ g NaOH} \times (1 \text{ mol NaOH}/40.0 \text{ g NaOH}) \times (1 \text{ mol OH}^-/1 \text{ mol NaOH}) = 1.4 \times 10^{-3} \text{ mol OH}^-$

    $[OH^-] = 1.4 \times 10^{-3} \text{ mol OH}^-/35 \, L = 3.9 \times 10^{-5} \, M \, OH^-$

    $pOH = -\log[OH^-] = -\log(3.9 \times 10^{-5} \, M \, OH^-) = 4.41$

    $14.00 = pH + pOH$

    $pH = 14.00 - pOH = 14.00 - 4.41 = 9.59$ **(pH and pOH)**

12. $K_a = ([H^+][F^-])/[HF]$ **(K_a expressions)**

13. $HC_4H_7O_2(aq) \rightleftarrows H^+(aq) + C_4H_7O_2^-(aq)$

Compound	Initial Concentration, M	Concentration Change, M	Equilibrium Concentration, M
$[HC_4H_7O_2]$	0.010	$-x$	$0.010 - x$
$[H^+]$	0.0	$+x$	x
$[C_4H_7O_2^-]$	0.0	$+x$	x

    $x = 0.010 \, HC_4H_7O_2 \times (3.9 \text{ molecules ionize}/100 \text{ molecules}) = 3.9 \times 10^{-4} \, M$

    $K_a = ([H^+][C_4H_7O_2^-])/[HC_4H_7O_2] =$

    $K_a = (3.9 \times 10^{-4} \, M)(3.9 \times 10^{-4} \, M)/0.010 \, M = 1.5 \times 10^{-5}$ **(K_a expressions)**

14. $HOCl(aq) \rightleftharpoons H^+(aq) + OCl^-(aq)$

   $pH = -\log [H^+]$

   $4.03 = -\log [H^+]$

   antilog$(-4.03) = [H^+]$

   $[H^+] = 9.3 \times 10^{-5} M$

   $[H^+] = [OCl^-] = 9.3 \times 10^{-5}$

   $[HOCl]_{initial} - [H^+] = [HOCl]_{final}$

   $0.25 M \text{ HOCl} - 0.000093 M = 0.25 M \text{ HOCl}$

   $K_a = ([H^+][OCl^-])/[HOCl] = ((9.3 \times 10^{-5} M)(9.3 \times 10^{-5} M))/0.25 M$

   $K_a = 3.5 \times 10^{-8}$ (K_a **values**)

15. $HC_2H_3O_2(aq) \rightleftharpoons H^+(aq) + C_2H_3O_2^-(aq)$

   $[HC_2H_3O_2] = 1.0 - x \quad [H^+] = x \quad [C_2H_3O_2^-] = x$

   $K_a = ([H^+][C_2H_3O_2^-])/[HC_2H_3O_2] = 1.8 \times 10^{-5}$

   $x^2/1.0 = 1.8 \times 10^{-5}$

   $x = 0.0042 M$

   $\% = (0.0042 M/1.0 M) \times 100 = 0.42\%$ (**Percent ionization**)

16.

Compound	Initial Concentration, M	Concentration Change, M	Equilibrium Concentration, M
$[HC_3H_5O_3]$	0.0010	$-x$	$0.0010 - x$
$[H^+]$	0.0	$+x$	x
$[C_2H_5O_3^-]$	0.0	$+x$	x

$K_a = ([H^+][C_3H_5O_3^-])/[HC_3H_5O_3] = 1.4 \times 10^{-4}$

$x^2/0.0010 - x = 1.4 \times 10^{-4}$

$x^2 = 1.4 \times 10^{-4}(0.0010 - x) = 1.4 \times 10^{-7} - 1.4 \times 10^{-4}x$

$x^2 + 1.4 \times 10^{-4}x - 1.4 \times 10^{-7} = 0$

$x = [H^+] = [C_3H_5O_3^-] = 3.1 \times 10^{-4} M$

$[HC_3H_5O_3] = 0.0010 - x = 0.0010 M - 0.00031 M = 7 \times 10^{-4} M$

$pH = -\log [H^+] = -\log (3.1 \times 10^{-4} M) = 3.51$ (K_a **expressions**)

17. $HC_2H_2O_2Cl(aq) \rightleftarrows H^+(aq) + C_2H_2O_2Cl^-(aq)$

    $K_a = [H^+][C_2H_2O_2Cl^-]/[HC_2H_2O_2Cl] = 1.4 \times 10^{-3}$

    $x^2/0.30\ M - x = 1.4 \times 10^{-3}$

    $x = 0.020\ M$

    $pH = -\log [H^+] = -\log(0.020\ M) = 1.70$ **(pH)**

18. $K_a = ([H^+][F^-])/[HF]$

    $HF(aq) \rightleftarrows H^+(aq) + F^-(aq)$

Compound	Initial Concentration, M	Concentration Change, M	Equilibrium Concentration, M
[HF]	0.100	$-x$	$0.100 - x$
[H$^+$]	0.010	$+x$	$0.010 + x$
[F$^-$]	0.0	$+x$	x

    $K_a = (0.010 + x)x/0.100 - x = 7.4 \times 10^{-4}$

    $x = 0.0047\ M$

    $[HF] = 0.100\ M - 0.0047 = 0.095\ M$ **(Common ion effect)**

19. $H_2SO_4(aq)\ H^+(aq) + HSO_4^-(aq)$

    $HSO_4^-(aq) \rightleftarrows H^+(aq) + SO_4^{2-}(aq)$ **(Polyprotic acids)**

20. $H_3PO_4(aq) \rightleftarrows H^+(aq) + H_2SO_4^-(aq)$

    $H_2PO_4-(aq) \rightleftarrows H^+(aq) + HPO_4^{2-}(aq)$

    $HPO_4^{2-}(aq)\ H^+(aq) + PO_4^{3-}(aq)$ **(Polyprotic acids)**

21.

Compound	Initial Concentration, M	Concentration Change, M	Equilibrium Concentration, M
[H$_3$PO$_4$]	1.5	$-x$	$1.5 - x$
[H$^+$]	0.0	$+x$	x
[H$_2$PO$_4^-$]	0.0	$+x$	x

    $K_a = x^2/1.50 - x = 7.5 \times 10^{-3}$

    $x = [H^+] = [H_2PO_4^-] = 0.10\ M$

    $pH = -\log [H^+] = -\log(0.10\ M) = 1.00$ **(Polyprotic acids)**

22. $x = [H^+] = [H_2PO_4^-] = 0.10\ M$ **(Polyprotic acid)**

Acid-Base Equilibrium in Aqueous Solution / 135

23. $[H^+] = [H_2PO_4^-] = 0.10\ M$ because $H_2PO_4^-$ is a very weak acid.

    $K_{a2} = [H^+][HPO_4^{2-}]/[H_2PO_4^-] = 6.2 \times 10^{-8}$

    $K_{a2} = (0.10\ M)[HPO_4^{2-}]/(0.10\ M)$

    $[HPO_4^{2-}] = 6.2 \times 10^{-8}\ M$ **(Polyprotic acids)**

24. $[H^+] = 0.10\ M$ and $[HPO_4^{2-}] = 6.2 \times 10^{-8}\ M$

    $K_{a3} = [H^+][PO_4^{3-}]/[HPO_4^{2-}] = 4.8 \times 10^{-13}$

    $K_{a3} = (0.10\ M)\ [PO_4^{3-}]/6.2 \times 10^{-8}\ M$

    $[PO_4^{3-}] = 2.9 \times 10^{-19}\ M$ **(Polyprotic acids)**

25. a. $NH_3(aq) + H_2O(l) \rightleftarrows NH_4^+(aq) + OH^-(aq)$

    b. $K_b = [NH_4^+][OH^-]/[NH_3] = 1.8 \times 10^{-5}$ **(K_b expressions)**

26. $H_2NOH(aq) + H_2O(l) \rightleftarrows H_3NOH^+(aq) + OH^-(aq)$

Compound	Initial Concentration, M	Concentration Change, M	Equilibrium Concentration, M
$[H_2NOH]$	0.12	$-x$	$0.12 - x$
$[OH^-]$	0.0	$+x$	x
$[H_3NOH^+]$	0.0	$+x$	x

    $K_b = [H_3NOH^+][OH^-]/[H_2NOH] = 1.1 \times 10^{-8}$

    $K_b = x^2/0.12 - x$

    $x = [OH^-] = 3.6 \times 10^{-5}\ M$

    $pOH = -\log[OH^-] = -\log(3.6 \times 10^{-5}\ M) = 4.44$

    $pH = 14 - pOH = 14 - 4.44 = 9.56$ **(Base ionizations)**

27. $IO^-(aq) + H_2O(l) \rightleftarrows HIO(aq) + OH^-(aq)$

Compound	Initial Concentration, M	Concentration Change, M	Equilibrium Concentration, M
$[IO^-]$	0.030	$-x$	$0.030 - x$
$[OH^-]$	0.0	$+x$	x
$[HIO]$	0.0	$+x$	x

    $pH + pOH = 14$

    $pOH = 14 - pH = 14 - 11.56 = 2.44$

    $pOH = -\log[OH^-]$

    $2.44 = -\log[OH^-]$

    $[OH^-] = 3.6 \times 10^{-3}\ M$

$x = [OH^-] = 3.6 \times 10^{-3} \, M$

$x = [HOI] = 3.6 \times 10^{-3} \, M$

$[OI^-] = 0.030 \, M - 3.6 \times 10^{-3} \, M = 0.026 \, M$

$K_b = [HIO][OH^-]/[IO^-] = (3.6 \times 10^{-3})^2/0.026 = 5.0 \times 10^{-4}$

$K_a = K_w/K_b = 1.0 \times 10^{-14}/5.0 \times 10^{-4} = 2.0 \times 10^{-11}$ (K_a-K_b Relationship)

28. $K_w = K_a \times K_b$

    $-\log K_w = -\log K_a + (-\log K_b)$

    $pK_w = pK_a + pK_b$

    $14 = pK_a + pK_b$ (pK_a- pK_b Relationship)

29. $14.00 = pK_a + pK_b$

    $pK_b = 14.00 - pK_a = 14.00 - 8.60 = 5.40$

    $pK_b = -\log K_b = 5.40$

    antilog(log K_b) = antilog(-5.40)

    $K_b = 4.0 \times 10^{-6}$ (pK_a- pK_b Relationship)

30. $Ca(C_3H_5O_3)_2(s) \xrightarrow{H_2O} Ca^{2+}(aq) + 2C_3H_5O_3^-(aq)$

    $C_3H_5O_3^-(aq) + H_2O(l) \rightleftarrows HC_3H_5O_3(aq) + OH^-(aq)$

Compound	Initial Concentration, M	Concentration Change, M	Equilibrium Concentration, M
[$C_3H_5O_3^-$]	0.80	$-x$	$0.80 - x$
[OH^-]	0.0	$+x$	x
[$HC_3H_5O_3$]	0.0	$+x$	x

    $K_b = K_w/K_a = 1.0 \times 10^{-14}/1.4 \times 10^{-4} = 7.1 \times 10^{-11}$

    $K_b = [HC_3H_5O_3][OH^-]/[C_3H_5O_3^-]$

    $x^2/0.80 - x = 7.1 \times 10^{-11}$

    $x = [OH^-] = 7.5 \times 10^{-6} \, M$

    $pOH = -\log(7.5 \times 10^{-6} \, M) = 5.12$

    $pH = 14 - pOH = 14 - 5.12 = 8.88$ (pH of salt solutions)

31. $HCF_3COO \rightleftarrows H^+ + CF_3COO^-$

    $[H^+] = [CF_3COO^-] = 0.19 \, M \times (79.6/100) = 0.15 \, M$

    $[HCF_3COO] = 0.19 \, M - 0.15 \, M = 0.04 \, M$

    $K_a = [H^+][CF_3COO^-]/[HCF_3COO] = (0.15 \, M)^2/0.04 \, M = 0.6$ (K_a values)

32. $pH = 2.89 = -\log[H^+]$

    $[H^+] = 1.3 \times 10^{-3} \, M$

$C_6H_5COOH \rightarrow H^+ + C_6H_5COO^-$

$[H^+] = [C_6H_5COO^-]$

$K_a = [H^+][C_6H_5COO^-]/[HC_6H_5COO]$

$K_a = (1.3 \times 10^{-3}\ M)^2/[HC_6H_5COO] - 1.3 \times 10^{-3}\ M = 6.5 \times 10^{-5}$

$[HC_6H_5COO] = 0.027\ M$ (K_a expressions)

33. $2NH_3 \rightleftarrows NH_4^+ + NH_2^-$ (Self ionization)

34. $HF \rightleftarrows H^+ + F^-$

    $[F^-] = 0.10\ M,\ [HF] = 0.015\ mol/1.00\ L = 0.015\ M$

    Because of the F^- ion concentration, the common ion, the value of x will be small enough to drop.

    $K_a = [H^+][F^-]/[HF] = 7.2 \times 10^{-4}$

    $(x \times 0.10\ M) / (0.015\ M - x) = 7.2 \times 10^{-4}$

    $x = [H^+] = 1.1 \times 10^{-4}\ M$ (Common ion effect)

35. c. $[H^+] = 0.10\ mol/10\ L = 0.010\ M\ H^+$

    $pH = -\log[H^+] = -\log(0.010\ M) = 2.00$ (pH)

36. d.

    $HF(aq) \rightleftarrows H^+(aq) + F^-(aq)$

    $x = [H^+] = [F^-]$

    $K_a = x^2/0.100 - x = 7.2 \times 10^{-4}$

    $x = [H^+] = 0.0081\ M$ (Hydrogen ion concentration)

37. b.

    mol $OH^- = 4.0\ g\ NaOH \times 1\ mol\ NaOH/40\ g = 0.10\ mol\ NaOH$

    $[OH^-] = 0.10\ mol\ NaOH/ 20.0\ L = 5.0 \times 10^{-3}\ M\ OH^-$

    $[H^+] = K_w/[OH^-] = 1.0 \times 10^{-14}/5.0 \times 10^{-3}\ M\ OH^- = 2.0 \times 10^{-12}\ M\ OH^-$ (Water self ionization)

38. b. $C_5H_5N + H_2O \rightleftarrows OH^- + C_5H_6N^+$

    $K_b = [C_5H_6N] + [OH^-]/[C_5H_5N]$

    $x^2/(0.75 - x) = 1.4 \times 10^{-9}$

    $x = [OH^-] = 3.2 \times 10^{-5}\ M$

    $pOH = -\log(3.2 \times 10^{-5}\ M) = 4.49$

    $pH = 14 - pOH = 14 - 4.49 = 9.51$ (K_b expressions)

39. d. $K_w = K_a K_b$

    $K_b = K_w/K_a = 1.0 \times 10^{-14}/1.8 \times 10^{-5} = 5.6 \times 10^{-10}$ (K_a-K_b Relationship)

40. d. $pOH = -\log(1.0 \times 10^{-10}\ M) = 10.00$

    $pH = 14 - 10.00 = 4.00$ (pH and pOH)

Grade Yourself

Circle the numbers of the questions you missed, then fill in the total incorrect for each topic. If you answered more than three questions incorrectly, you need to focus on that topic. (If a topic has less than three questions and you had at least one wrong, we suggest you study that topic also. Read your textbook, a review book, or ask your teacher for help.)

Subject: Acid-Base Equilibrium in Aqueous Solution

Topic	Question Numbers	Number Incorrect
Water self-ionization	1, 2, 37	
H^+ and OH^- concentrations	3, 4, 5,	
pH	6, 7, 8, 9, 11, 17, 35, 40	
pOH	10, 11, 40	
K_a expressions	12, 13, 32,	
K_a values	14, 16, 31,	
K_b expressions	25, 38	
K_a - K_b relationship	27, 39	
pK_a - pK_b relationship	28, 29,	
Percent ionization	15,	
Polyprotic acids	19, 20, 21, 22, 23, 24,	
Base ionizations	26,	
pH of salt solutions	30,	
Self ionization	33	
Common ion effect	18, 34	
Hydrogen ion concentration	36	

Applications of Aqueous Equilibria

 Brief Yourself

Buffer solutions are those that resist changes in pH. They consist of either a weak acid and its conjugate base or a weak base and its conjugate acid. The basic component of the buffer neutralizes small amounts of added acids and the acidic component neutralizes small amounts of added bases.

The Henderson-Hasselbalch equation is often used to solve buffer solution problems. This equation is as follows:

$$pH = pK_a + \log\frac{[A^-]}{[HA]}$$

In this equation, $[A^-]$ is the molar concentration of the conjugate base and $[HA]$ is the molar concentration of a weak acid. The Henderson-Hasselbalch equation is used to calculate the pH of a buffer from the pK_a and the common logarithm of the ratio of the molar concentrations of the conjugate base and weak acid. Each buffer solution has a fixed capacity to maintain a constant pH. The buffer capacity of a solution depends on the concentration of the components and how close the molar ratio of the components is to 1.

An acid-base titration is the systematic addition of a base (or acid) of known concentration to an acid (or base) of unknown concentration. A titration curve is a plot of the pH of the solution versus the volume of the added acid or base. The shape of a titration curve depends on the general type of acid and base titrated. In the titration of a strong acid by a strong base the titration curve begins at a low pH value because of the presence of a strong acid. Initially, the addition of base produces a small change in pH. Not until just before the equivalence point, pH 7, is reached does the pH change significantly. Addition of a drop of strong base near the equivalence point changes the pH by as much as five to six pH units. Beyond the equivalence point the pH initially rises rapidly and then levels off.

The titration curve for a weak acid and strong base differs noticeably from that of a strong acid and strong base. The starting pH is higher than that of a strong acid-strong base titration because of the weaker acid. The pH curve rises rapidly with the addition of the first few milliliters of base, and then levels off through the buffer region where the concentration of the weak acid is approximately equal to that of the conjugate base. The increase in pH near the equivalence point is much less than that in a strong acid and base, rising only 3 to 4 pH units instead of 5 to 6 pH units. Finally, the equivalence point occurs at a pH above 7.

The titration curve for a weak base and strong acid resembles that of a weak acid and strong base, but it decreases in pH with the addition of strong acid. The starting pH is always high because a weak base is present initially. The curve drops rapidly with the initial addition of acid but levels off when a significant amount of conjugate acid is present–a buffer solution results. The pH change through the equivalence point is about the same as that in a weak acid-strong base titration. The equivalence point is always at a pH below 7 because the principal species present is the conjugate acid of a weak base.

The equilibrium that establishes between a partially soluble ionic solid and its ions is a solubility equilibrium. This equilibrium is describe using the solubility-product equilibrium expression. The product of the molar concentrations of the dissolved ions raised to the power equal to their coefficients in the equation equals the solubility-product equilibrium constant, K_{sp}. This equilibrium constant allows us to calculate the solubility of solids. However, the K_{sp} is not the solubility of a solid. The K_{sp} values can be used to compare solubilities only if the solid dissociates in a similar manner and produces the same number of dissolved ions in solution. The addition of a common ion shifts solubility equilibria to the left, which makes the solids less soluble. Removal of an ion causes more solid to dissolve.

Solubility product equilibrium expressions can be used to predict if a precipitate forms in a precipitation reaction. If the ion product, Q, is less than K_{sp} then no precipitate forms because the concentration of the ions is too small to establish equilibrium. If Q equals K_{sp} then the solution just becomes saturated, and when Q exceeds K_{sp} then a precipitate falls from solution. Chemists use selective precipitation to separate ions. For example, this is accomplished by selecting an anion that precipitates one cation, leaving the other in solution. A qualitative inorganic analysis scheme makes use of the differences in solubility of inorganic ions to separate them into five groups.

Complex ions result when Lewis bases, ligands, combine with metal ions. Complex ions can be used to change the solubility of partially soluble solids. A ligand bonds to dissolved metal ions and forms soluble complex ions. This removes the metal ions in equilibrium with the insoluble solid and shifts the equilibrium to the right, towards the dissolved ions. Complex ions are in equilibrium with the ions they dissociate to. Thus, an equilibrium expression can be written. Chemists either use formation equilibrium constants, K_f, in which the equilibrium has the complex ion as product, or the dissociation constant, K_d, in which the complex ion is a reactant.

Test Yourself

1. a. What is a buffer solution? b. How is a buffer solution prepared?

2. a. Write an equation for the equilibrium that is established in an acetic acid-acetate buffer. b. Write an equation that shows what happens when acid, H^+, is added to this buffer. c. Write an equation that shows what happens when base, OH^-, is added to this buffer.

3. a. Write an equation for the equilibrium that establishes in an ammonia-ammonium ion buffer. b. Write equations that show what happens when either acid or base is added to this buffer.

For Problems 4 to 6, consider 100 mL of a buffer solution that contains 0.10 M $HC_2H_3O_2$ and 0.10 M $C_2H_3O_2^-$. The K_a for $HC_2H_3O_2$ is 1.8×10^{-5}.

4. What is the pH of this buffer solution?

5. Calculate the pH after adding 0.0010 mol of a strong solid acid to 100 mL of this buffer.

6. Calculate the pH after adding 0.0010 mol of a NaOH(s) to 100 mL of this buffer.

For Problems 7 and 8, consider 250 mL of a buffer solution that contains 0.500 M HOCl and 0.350 M NaOCl. The K_a for HOCl is 3.5×10^{-8}.

7. What is the pH of this buffer solution?

8. What is the pH of the solution after 1.0 mL 1.0 M NaOH is added?

9. Given 0.50 M $HC_2H_3O_2$ and 0.50 M $NaC_2H_3O_2$ explain how 1.0 L of a buffer solution is prepared with a pH of 4.60. The K_a for acetic acid is 1.8×10^{-5}.

10. a. What is the Henderson-Hasselbalch equation? b. How is it used? c. Use this equation, to determine what the pH of a solution is when the conjugate base and weak acid concentration is equal.

11. A buffer solution contains 4.6 g sodium propanoate, $NaC_3H_5O_2(s)$, in 165 mL of 0.36 M propanoic acid, $HC_3H_5O_2$. The K_a for propanoic acid is 1.3×10^{-5}. Use the Henderson-Hasselbalch equation to calculate the pH of this buffer. Make the assumption that the addition of the solid sodium propanoate does not change the total volume of the solution.

For Problems 12 and 13 consider 100 mL of a buffer that contains 0.550 M NH_3 and 0.650 M NH_4Cl. The K_b for NH_3 is 1.8×10^{-5}.

12. What is the pH of this buffer?

13. What is the pH of the buffer after the addition of 0.20 g NaOH(s)? Assume that the solid NaOH does not change the total volume of the buffer solution.

14. a. What is buffer capacity? b. What factors determine the buffer capacity of a solution?

15. a. Write the net ionic equation when a strong acid is neutralized by a strong base. b. What is the relationship of the moles of H^+ and OH^- at the equivalence point in the titration of a strong acid and strong base? c. What acid-base indicator is used in such a titration?

For Problems 16 to 20, consider the titration of 25.00 mL 0.1000 M HCl with 0.1000 M NaOH.

16. What is the initial pH of the solution before any NaOH is added?

17. What is the pH of the solution after 15.00 mL 0.1000 M NaOH is added to the 25.00 mL 0.1000 M HCl?

18. What is the pH of the solution after 24.00 mL 0.1000 M NaOH is added to the 25.00 mL 0.1000 M HCl?

19. What is the pH of this solution at the equivalence point?

20. Calculate the pH after the addition of 26.00 mL 0.1000 M NaOH.

For Problems 21 to 24, consider the titration of 25.00 mL 0.1000 M $HC_2H_3O_2$ with 0.1000 M NaOH. The K_a of $HC_2H_3O_2$ is 1.8×10^{-5}.

21. What is the initial pH of the solution?

22. What is the pH of the solution after 10.00 mL 0.1000 NaOH is added to 25.00 mL 0.1000 M $HC_2H_3O_2$?

23. What is the pH of the solution at the half neutralization point?

24. What is the pH of the solution at the equivalence point?

For Problems 25 to 27, consider the titration of 25.00 mL of 0.1000 M NH$_3$ with 0.1000 M HCl. The K_b for NH$_3$ is 1.8×10^{-5}.

25. What is the pH after 10.00 mL 0.1000 M HCl is added to 25.00 mL 0.1000 NH$_3$?

26. What is the pH at the half-neutralization point?

27. What is the pH at the equivalence point?

For Problems 28 and 29, consider the solubility equilibrium established by barium sulfate.

28. a. Write the equation for the solubility equilibrium established by barium sulfate.
    b. Write the K_{sp} expression for barium sulfate.

29. At 298 K, the molar solubility of barium sulfate is 4×10^{-5} mol/L. What is the K_{sp} value for barium sulfate?

30. a. Write the equation for the solubility equilibrium of calcium phosphate. b. Write the K_{sp} for calcium phosphate.

31. Explain the difference between the K_{sp} value and the solubility of a partially soluble substance.

32. A 0.384-g sample of PbBr$_2$ is required to just saturate 100 mL of water at 298 K. Calculate the K_{sp} value for PbBr$_2$ at 298 K.

33. Calculate the molar solubility of Ag$_2$CrO$_4$. The K_{sp} value for silver chromate is 9.0×10^{-12}.

34. Will a precipitate form when 10 mL 0.030 M Pb(NO$_3$)$_2$ is mixed with 20 mL 0.0060 M NaCl? The K_{sp} value for PbCl$_2$ is 1.6×10^{-5}.

35. What is the pH of the resulting solution after 50.0 mL 0.150 M MgCl$_2$ is added to 50. mL 0.300 M NaOH? The K_{sp} of Mg(OH)$_2$ is 1.2×10^{-11}.
    a. 13.48
    b. 12.88
    c. 10.45
    d. 10.92
    e. none of these

36. Calculate the pH of a ammonia-ammonium ion buffer solution that is prepared by mixing 150 mL of 0.88 M NH$_3$ and 150 mL of 0.66 M NH$_4$Cl. Assume the volumes are additive. The K_b for NH$_3$ is 1.8×10^{-5}.
    a. 9.26
    b. 9.14
    c. 9.38
    d. 9.52
    e. none of these

37. If 75 mL 0.045 M AgNO$_3$ mixes with 125 mL 1.00 M HCl, what percent of Ag$^+$ remains in solution after the addition of the HCl? The K_{sp} of AgCl is 1.6×10^{-10}.
    a. 1.6×10^{-4} %
    b. 1.6×10^{-10} %
    c. 2.6×10^{-10} %
    d. 5.8×10^{-7} %
    e. none of these

38. Which of the following conditions could be used to precipitate the qualitative analysis group III cations?
    a. low pH and high chloride ion concentration
    b. low pH and low sulfide ion concentration
    c. high pH and high sulfide ion concentration
    d. high pH and high carbonate concentration
    e. none of these

39. Under what conditions would Hg^{2+} be precipitated in a qualitative analysis scheme?
    a. low pH and high chloride ion concentration
    b. low pH and low sulfide ion concentration
    c. high pH and high sulfide ion concentration
    d. high pH and high carbonate concentration
    e. none of these

40. Which of the following cations belongs to qualitative analysis group V?
    a. Ag$^+$
    b. Al^{3+}
    c. Ca^{2+}
    d. Mg^{2+}
    e. none of these

Check Yourself

1. a. Buffer solutions maintain a nearly constant pH with the addition of small amounts of either acids or bases. b. Buffer solutions are prepared by mixing either a weak acid and its conjugate base, or a weak base and its conjugate acid. **(Buffer solutions)**

2. a. $HC_2H_3O_2(aq) \rightleftarrows H^+(aq) + C_2H_3O_2^-(aq)$

   b. $H^+(aq) + C_2H_3O_2^-(aq) \rightarrow HC_2H_3O_2(aq)$ (added acid)

   $OH^-(aq) + HC_2H_3O_2(aq) \rightarrow C_2H_3O_2^-(aq) + H_2O(l)$ (added base)   **(Buffer solutions)**

3. a. $NH_3(aq) + H_2O(l) \rightleftarrows NH_4^+(aq) + OH^-(aq)$

   b. $H^+(aq) + NH_3(aq) \rightarrow NH_4^+(aq)$

   $OH^-(aq) + NH_4^+(aq) \rightarrow NH_3(aq) + H_2O(l)$ **(Buffer solutions)**

4. $K_a = [H^+][C_2H_3O_2^-]/[HC_2H_3O_2] = 1.8 \times 10^{-5}$

   $[H^+](0.100\ M)/(0.100\ M) = 1.8 \times 10^{-5}$

   $[H^+] = 1.8 \times 10^{-5}$

   $pH = -\log[H^+] = -\log(1.8 \times 10^{-5}\ M) = 4.74$ **(Buffer solutions)**

5. $H^+(aq) + C_2H_3O_2^-(aq) \rightarrow HC_2H_3O_2(aq)$

   mol $HC_2H_3O_2$ = mol $C_2H_3O_2^-$ = 0.100 L × 0.100 mol/L = 0.0100 mol

   mol $C_2H_3O_2^-$ = 0.0100 mol – 0.0010 mol = 0.0090 mol $C_2H_3O_2^-$

   mol $HC_2H_3O_2$ = 0.0100 mol + 0.0010 mol = 0.0110 mol $HC_2H_3O_2$

   $[C_2H_3O_2^-]$ = 0.0090 mol $C_2H_3O_2^-$/0.100 L = 0.090 M

   $[HC_2H_3O_2]$ = 0.0110 mol $HC_2H_3O_2$/0.100 L = 0.110 M

   $K_a = [H^+][C_2H_3O_2^-]/[HC_2H_3O_2] = 1.8 \times 10^{-5}$

   $[H^+](0.090\ M)/(0.110\ M) = 1.8 \times 10^{-5}$

   $[H^+] = 2.2 \times 10^{-5}\ M$

   $pH = -\log[H^+] = -\log(2.2 \times 10^{-5}\ M) = 4.66$ **(Buffer solutions)**

6. $OH^-(aq) + HC_2H_3O_2(aq) \rightarrow C_2H_3O_2^-(aq) + H_2O(l)$

   mol $HC_2H_3O_2$ = mol $C_2H_3O_2^-$ = 0.0100 mol

   mol $C_2H_3O_2^-$ = 0.0100 mol + 0.0010 mol = 0.0110 mol $C_2H_3O_2^-$

   $[C_2H_3O_2^-]$ = 0.0110 mol $HC_2H_3O_2$/0.100 L = 0.110 M

   mol $HC_2H_3O_2$ = 0.0100 mol – 0.0010 mol = 0.0090 mol $HC_2H_3O_2$

   $[HC_2H_3O_2]$ = 0.0090 mol $C_2H_3O_2^-$/0.100 L = 0.090 M

   $K_a = [H^+][C_2H_3O_2-]/[HC_2H_3O_2] = 1.8 \times 10^{-5}$

   $[H^+](0.110\ M)/(0.090\ M) = 1.8 \times 10^{-5}$

   $[H^+] = 1.5 \times 10^{-5}\ M$

   $pH = -\log[H^+] = -\log(1.5 \times 10^{-5}) = 4.83$ **(Buffer solutions)**

7. $HOCl(aq) \rightleftarrows H^+(aq) + OCl^-$

$K_a = [H^+][OCl^-]/[HOCl] = 3.5 \times 10^{-8}$

Compound	Initial Concentration, M	Concentration Change, M	Equilibrium Concentration, M
[HOCl]	0.500	$-x$	$0.500 - x$
[H$^+$]	0.0	$+x$	x
[OCl$^-$]	0.350	$+x$	$0.350 + x$

$x\,(0.350\,M + x)/(0.500\,M - x) = 3.5 \times 10^{-8}$

$x\,(0.350\,M)/(0.500\,M) = 3.5 \times 10^{-8}$

$x = [H^+] = 5.0 \times 10^{-8}\,M$

$pH = -\log[H^+] = -\log(5.0 \times 10^{-8}\,M) = 7.30$ **(Buffer solutions)**

8. $OH^-(aq) + HOCl(aq) \rightarrow H_2O(l) + OCl^-(aq)$

   mol OH^- = 1.0 mL $OH^- \times$ (1 L/1000 mL) $\times$ (1.0 mol OH^-) = 1.0×10^{-3} mol OH^- = 0.0010 mol OH^-

   mol HOCl = 0.250 L $\times$ 0.500 mol HOCl/L = 0.125 mol HOCl

   mol OCl^- = 0.250 L $\times$ 0.350 mol OCl^-/L = 0.0875 mol OCl^-

   mol HOCl = 0.125 mol − 0.0010 mol = 0.124 mol HOCl

   mol OCl^- = 0.0875 mol + 0.0010 mol = 0.0885 mol OCl^-

   [HOCl] = 0.124 mol HOCl/0.251 L = 0.494 M HOCl

   [OCl^-] = 0.0885 mol OCl^-/0.251 L = 0.353 M OCl^-

   $K_a = [H^+][OCl^-]/[HOCl] = 3.5 \times 10^{-8}$

   $[H^+](0.353\,M)/(0.494\,M) = 3.5 \times 10^{-8}$

   $[H^+] = 4.9 \times 10^{-8}\,M$

   $pH = -\log[H^+] = -\log(4.9 \times 10^{-8}\,M) = 7.31$ **(Buffer solutions)**

9. $HC_2H_3O_2(aq) \rightleftarrows H^+(aq) + C_2H_3O_2^-(aq)$

   $pH = -\log[H^+]$

   $4.60 = -\log[H^+]$

   $[H^+] = 2.5 \times 10^{-5}\,M$

   $K_a = [H^+][C_2H_3O_2^-]/[HC_2H_3O_2] = 1.8 \times 10^{-5}$

   $[C_2H_3O_2^-]/[HC_2H_3O_2] = 1.8 \times 10^{-5}/[H^+] = 1.8 \times 10^{-5}/2.5 \times 10^{-5}\,M = 0.72$

   If x equals the volume of acetate, then 1.00 − x is the volume of acetic acid because the total volume is 1.00 L. Thus, the volume ratio is expressed as follows.

   $$x/1.00 - x = 0.72$$

   Solving this equation for x yields 0.42 L. This means that 0.42 L of acetate is required and 1.00 L − 0.42 L or 0.58 L of acetic acid is required to prepare 1.00 L of a buffer solution with a pH of 4.60. **(Buffer solution)**

10. a. $pH = pK_a + \log([A^-]/[HA])$, b. It is used to simplify buffer solution calculations., c. When [A^-] equals [HA], the pH of the solution equals pK_a. **(Buffer solutions)**

11. $HC_3H_5O_2(aq) \rightleftarrows H^+(aq) + C_3H_5O_2^-(aq)$

pH = pK_a + log([$C_3H_5O_2^-$]/[$HC_3H_5O_2$])

pK_a = –log K_a = –log (1.3 × 10^{-5}) = 4.89

mol $C_3H_5O_2^-$ = 4.6 g $NaC_3H_5O_2$ × (1 mol $NaC_3H_5O_2$/96.0 g $NaC_3H_5O_2$) ×
$\qquad\qquad\qquad\qquad$ (1 mol $C_3H_5O_2^-$/1 mol $NaC_3H_5O_2$) = 0.048 mol $C_3H_5O_2^-$

[$C_3H_5O_2^-$] = 0.048 mol $C_3H_5O_2^-$/0.165 L = 0.29 M $C_3H_5O_2^-$

pH = pK_a + log ([$C_3H_5O_2^-$]/[$HC_3H_5O_2$])

pH = 4.89 + log (0.29 M $C_3H_5O_2^-$/0.36 M $HC_3H_5O_2$) = 4.80 **(Buffer solutions)**

12. $NH_3(aq)$ + $H_2O(l)$ ⇌ $NH_4^+(aq)$ + $OH^-(aq)$

    pK_b = –log K_b = –log (1.8 × 10^{-5}) = 4.74

    pOH = pK_b + log ([NH_4^+]/[NH_3]) = 4.74 + log (0.650 M NH_4^+/0.550 M NH_3) = 4.81

    pH + pOH = 14

    pH = 14 – pOH = 14 – 4.81 = 9.19 **(Buffer solutions)**

13. $OH^-(aq)$ + $NH_4^+(aq)$ ⇌ $H_2O(l)$ + $NH_3(aq)$

    mol OH^- = 0.20 g NaOH × (1 mol NaOH/40.0 g NaOH) × (1 mol OH^-/1 mol NaOH) = 0.0050 mol OH^-

    mol NH_3 = 100 mL × (1 L/1000 mL) × (0.550 mol NH_3/L) = 0.0550 mol NH_3

    mol NH_4^+ = 100 mL × (1 L/1000 mL) × (0.650 mol NH_4Cl/L) × (1 mol NH_4^+/1 mol NH_4Cl)
    $\qquad\qquad\qquad\qquad\qquad\qquad\qquad\qquad\qquad\qquad\qquad$ = 0.0650 mol NH_4^+

    mol NH_4^+ = 0.0650 mol NH_4^+ – 0.0050 mol = 0.0600 mol NH_4^+

    mol NH_3 = 0.0550 mol NH_3 + 0.0050 mol = 0.0600 mol NH_3

    pOH = pK_b + log ([NH_4^+]/[NH_3]) = 4.74 + log (0.600 M NH_4^+/0.600 M NH_3) = 4.74

    pH = 14 – 4.74 = 9.26 **(Buffer solutions)**

14. a. Buffer capacity refers to the amount of acid or base that may be added to a buffer before the pH changes significantly. b. Buffer capacity depends on the concentration of the acidic and basic components. The higher the concentration of these components and the closer their ratio is to 1, the greater the buffer capacity. **(Buffer capacity)**

15. a. $H^+(aq)$ + $OH^-(aq)$ → $H_2O(l)$, b. mol H^+ = mol OH^-, c. phenolphthalein . **(Titrations)**

16. pH = –log (0.1000 M HCl) = 1.0000   **(Titrations)**

17. 15.00 mL 0.1000 M NaOH neutralizes 15.00 mL 0.1000 M HCl, leaving 10.00 mL of excess HCl solution

    mol H^+ = 10.00 mL × (1L/1000 mL) × (0.1000 mol H^+/L) = 1.000 × 10^{-3} mol H^+

    [H^+] = 1.000 × 10^{-3} mol H^+/0.04000 L = 2.500 × 10^{-2} M H^+

    pH = –log (2.500 × 10^{-2} M HCl) = 1.6021 **(Titrations)**

18. Only 1.00 mL of excess 0.1000 M HCl remains at this point.

    mol H^+ = 1.00 mL × 1L/1000 mL × 0.1000 mol H^+/L = 1.00 × 10^{-4} mol H^+

    [H^+] = 1.00 × 10^{-4} mol H^+/0.04900 L = 2.04 × 10^{-3} M H^+

    pH = –log (2.04 × 10^{-2} M HCl) = 2.690 **(Titrations)**

19. At the equivalence point the moles of H^+ and OH^- are equal which means that the pH if 7. This is only true when a strong acid is neutralized by a strong base. **(Titration)**

20. At this point, 1.00 mL of excess NaOH is in the solution.

   mol OH⁻ = 1.00 mL × 1L/1000 mL × 0.1000 mol OH⁻/L = 1.00×10^{-4} mol OH⁻

   [OH⁻] = 1.00×10^{-4} mol H⁺/0.05100 L = 1.96×10^{-3} M OH⁻

   pOH = $-\log(1.96 \times 10^{-3}$ M OH⁻) = 2.690

   pH = 14 − 2.690 = 11.290 **(Titration)**

21. K_a = [H⁺][C₂H₃O₂⁻]/[HC₂H₃O₂] = 1.8×10^{-5}

   $x^2/(0.1000\ M - x) = 1.8 \times 10^{-5}$

   [H⁺] = 1.3×10^{-3}

   pH = −log [H⁺] = −log(1.3×10^{-3} M) = 2.87 **(Titration)**

22. $HC_2H_3O_2(aq) + OH^-(aq) \rightarrow C_2H_3O_2^-(aq) + H_2O(l)$

   mol HC₂H₃O₂ = 25.00 mL × (1 L/1000 mL) × (0.1000 mol HC₂H₃O₂/L) = 0.002500 mol HC₂H₃O₂

   mol OH⁻ = 10.00 mL × (1 L/1000 mL) × (0.1000 mol NaOH/L) ×

   (1 mol OH⁻/1 mol NaOH) = 0.001000 mol OH⁻

   mol HC₂H₃O₂ (remaining) = mol HC₂H₃O₂ (initial) − mol HC₂H₃O₂ (consumed)

   mol HC₂H₃O₂ (remaining) = 0.002500 mol HC₂H₃O₂ − (0.001000 mol OH⁻ ×

   (1 mol HC₂H₃O₂ consumed/1 mol OH⁻)) = 0.001500 mol HC₂H₃O₂

   [HC₂H₃O₂] = 0.001500 mol HC₂H₃O₂/0.03500 L = 0.04286 M HC₂H₃O₂

   [C₂H₃O₂⁻] = 0.001000 mol C₂H₃O₂⁻/0.03500 L = 0.02857 M C₂H₃O₂⁻

Compound	Initial Moles	Mole Change	Equilibrium Moles
HC₂H₃O₂	0.002500	−0.001000	0.001500
OH⁻	0.001000	−0.001000	0
C₂H₃O₂⁻	0.0	+0.001000	0.001000

   K_a = [H⁺][C₂H₃O₂⁻]/[HC₂H₃O₂] = 1.8×10^{-5}

   K_a = [H⁺] (0.02857 M)/ 0.04286 M

   [H⁺] = 2.7×10^{-5} M

   pH = −log(2.7×10^{-5} M) = 4.57 **(Titration)**

23. Knowing that 25 mL 0.1000 M NaOH neutralizes the 25.00 mL HC₂H₃O₂, 12.50 mL are needed to half neutralize the solution. At this point the moles of HC₂H₃O₂ that remain equal the moles of C₂H₃O₂⁻ produced ([HC₂H₃O₂] = [C₂H₃O₂⁻]).

   With equal number of moles of acetic acid and acetate ions, the [H⁺] of the solution equals K_a and the pH equals pK_a.

   K_a = [H⁺][C₂H₃O₂⁻]/[HC₂H₃O₂] = 1.8×10^{-5}

$[H^+] = 1.8 \times 10^{-5}\ M$

$pH = -\log [H^+] = -\log K_a = pK_a = -\log (1.8 \times 10^{-5}\ M) = 4.74$ **(Titration)**

24. At the equivalence point the moles of OH^- added equal the moles of $HC_2H_3O_2$ originally present.

    mol OH^- = mol $HC_2H_3O_2$

    $HC_2H_3O_2(aq) + OH^-(aq) \rightarrow C_2H_3O_2^-(aq) + H_2O(l)$

    As soon as the $C_2H_3O_2^-$ forms, being a basic anion, it accepts protons from water and establishes the following equilibrium.

    $C_2H_3O_2^-(aq) + H_2O(l) \rightleftarrows HC_2H_3O_2(aq) + OH^-(aq)$

    This is the equilibrium for the base ionization of acetate and is represented by the base-ionization equilibrium expression.

    $K_b = K_w/K_a = 1.0 \times 10^{-14}/1.8 \times 10^{-5} = 5.6 \times 10^{-10}$

    $K_b = [HC_2H_3O_2][OH^-]/[C_2H_3O_2^-] = 5.6 \times 10^{-10}$

    $[C_2H_3O_2^-] = 0.002500$ mol $C_2H_3O_2^-/0.05000$ L $= 5.000 \times 10^{-2}\ M\ C_2H_3O_2^-$

    $K_b = x^2/5.000 \times 10^{-2} - x = 5.6 \times 10^{-10}$

    $x = [OH^-] = 5.3 \times 10^{-6}\ M$

    $pOH = -\log(5.3 \times 10^{-6}\ M) = 5.28.$

    $pH = 14 - 5.28 = 8.72$ **(Titration)**

25. $H^+(aq) + NH_3(aq) \rightarrow NH_4^+(aq)$

    moles of H^+ = 10.00 mL × (1 L/1000 mL) × (0.1000 mol HCl/L) × (1 mol H^+/1 mol HCl) =
    $$0.001000\ \text{mol}\ H^+$$

    and NH_3 = 25.00 mL × (1 L/1000 mL) × (0.1000 mol NH_3/L) = 0.002500 mol

    mol NH_3 (remaining) = mol NH_3 (initial) − mol H^+ = 0.002500 mol NH_3 − 0.001000 mol H^+ =
    $$0.001500\ \text{mol}\ NH_3$$

    For each one mole of H^+ neutralized, one mole of NH_4^+ results; thus, the moles of NH_4^+ produced is 0.001000 mol.

    $[NH_3]$ = 0.001500 mol NH_3/0.03500 L = 0.04286 M NH_3

    $[NH_4^+]$ = 0.001000 mol NH_4^+/0.03500 L = 0.02857 M NH_4^+

    $NH_3(aq) + H_2O(l) \rightleftarrows NH_4^+(aq) + OH^-(aq)$

    $K_b = 1.8 \times 10^{-5} = [NH_4^+][OH^-]/[NH_3] = (0.02857\ M)\ [OH^-]/(0.04286\ M)$

    $[OH^-] = 2.7 \times 10^{-5}\ M$

    $pOH = -\log (2.7 \times 10^{-5}\ M) = 4.57$

    $pH = 14 - pOH = 14 - 4.57 = 9.43$ **(Titration)**

148 / College Chemistry

26. The [NH$_3$] and [NH$_4^+$] are equal and divide out of the equilibrium equation ([NH$_3$] = [NH$_4^+$]) at the half-neutralization point.

    K_b = [NH$_4^+$][OH$^-$]/[NH$_3$] = 1.8×10^{-5}

    [OH$^-$] = 1.8×10^{-5} M

    pOH = pK_b = 4.74

    pH = 14 − 4.74 = 9.26   **(Titration)**

27. At the neutralization point, only NH$_4^+$ is present.

    NH$_4^+$(aq) → H$^+$(aq) + NH$_3$(aq)

    K_a = [H$^+$][NH$_3$]/[NH$_4^+$] = 5.56×10^{-10}

    [NH$_4^+$] = 0.002500 mol NH$_4^+$/0.05000 L = 5.000×10^{-2} M NH$_4^+$

    x = molarity of NH$_4^+$ that dissociates

    $K_a = x^2/(5.000 \times 10^{-2} - x) = 5.56 \times 10^{-10}$

    x = [H$^+$] = 5.3×10^{-6} M

    pH = −log (5.3×10^{-6} M) = 5.28   **(Titration)**

28. a. BaSO$_4$(s) $\underset{\leftarrow}{\overset{H_2O}{\rightarrow}}$ Ba^{2+}(aq) + SO$_4^{2-}$(aq)

    b. K_{sp} = [Ba^{2+}][SO$_4^{2-}$]   **(Solubility equilibria)**

29. [Ba^{2+}] = [SO$_4^{2-}$] = 4×10^{-5} mol/L

    K_{sp} = [Ba^{2+}][SO$_4^{2-}$] = $(4 \times 10^{-5}$ mol/L$)^2$ = 2×10^{-9} **(Solubility equilibria)**

30. a. Ca$_3$(PO$_4$)$_2$(aq) ⇌ 3Ca^{2+}(aq) + 2PO$_4^{3-}$(aq),

    b. K_{sp} = [Ca^{2+}]^3 [PO$_4^{3-}$]^2 **(Solubility equilibria)**

31. K_{sp} is the solubility-product equilibrium constant. It is not the molar solubility of a solid. The solubility of solids is measured in terms of the mass or moles of solid that dissolves in a fixed quantity of solvent. K_{sp} is a useful quantity because it has a constant value when the temperature is constant. Whereas the solubility of a substance can have many different values at a constant temperature. **(Solubility equilibria)**

32. PbBr$_2$(s) ⇌ Pb^{2+}(aq) + 2Br$^-$(aq)

    K_{sp} = [Pb^{2+}] [Br$^-$]^2

    0.384 g PbBr$_2$/100 mL × (1 mol PbBr$_2$/367 g PbBr$_2$) × (1000 mL/1 L) = 1.05×10^{-2} mol PbBr$_2$/L

    [Pb^{2+}] = 1.05×10^{-2} mol PbBr$_2$/L × (1 mol Pb^{2+}/1 mol PbBr$_2$) = 1.05×10^{-2} M

    [Br$^-$] = 1.05×10^{-2} mol PbBr$_2$/L × (2 mol Br$^-$/1 mol PbBr$_2$) = 2.10×10^{-2} M

    K_{sp} = [Pb^{2+}] [Br$^-$]^2 = $(1.05 \times 10^{-2}$ M$)(2.10 \times 10^{-2}$ M$)^2$ = 4.63×10^{-6} **(Solubility equilibria)**

33. Ag$_2$CrO$_4$(s) ⇌ 2Ag$^+$(aq) + CrO$_4^{2-}$(aq)

    [Ag$^+$]^2 [CrO$_4^{2-}$] = 9.0×10^{-12}

    x = M (CrO$_4^{2-}$) = molar solubility of Ag$_2$CrO$_4$

    $(2x)^2 x = 9.0 \times 10^{-12}$

    $4x^3 = 9.0 \times 10^{-12}$

    x = [CrO$_4^{2-}$] = 1.3×10^{-4} M   **(Solubility equilibria)**

34. $Pb(NO_3)_2(aq) + 2NaCl(aq) \rightleftarrows PbCl_2(s) + 2NaNO_3(aq)$

    $K_{sp} = [Pb^{2+}][Cl^-]^2$

    $Q = [Pb^{2+}][Cl^-]^2$

    $[Pb^{2+}]$ = 10 mL/30 mL $\times$ 0.030 M $Pb(NO_3)_2 \times$ (1 mol Pb^{2+}/1 mol $Pb(NO_3)_2$) = 0.010 M Pb^{2+}

    $[Cl^-]$ = 20 mL/30 mL $\times$ 0.0060 M NaCl $\times$ (1 mol Cl^-/1 mol NaCl) = 0.0040 M Cl^-

    $Q = [Pb^{2+}][Cl^-]^2 = (0.010\ M)(0.0040\ M)^2 = 6.4 \times 10^{-8}$

    Because 6.4×10^{-8} is significantly less than the K_{sp} value, 1.6×10^{-5}, no precipitate forms. **(Solubility equilibrium)**

35. c. $MgCl_2(aq) + 2NaOH(aq) \rightarrow Mg(OH)_2(s) + 2NaCl(aq)$

    0.0500 L $\times$ (0.150 mol Mg^{2+}/L) = 0.00750 mol Mg^{2+}

    0.0500 L $\times$ (0.300 mol OH^-/L) = 0.0150 mol OH^-

    $Mg^{2+} + 2OH^- \rightarrow Mg(OH)_2$

    All of the initial moles of OH^- (0.0150 mol) precipitates and the following equilibrium establishes.

    $Mg(OH)_2 \rightleftarrows Mg^{2+} + 2OH^-$

    $K_{sp} = [Mg^{2+}][OH^-]^2 = 1.2 \times 10^{-11}$

    $x \times (2x)^2 = 1.2 \times 10^{-11}$

    $x = [Mg^{2+}] = 1.4 \times 10^{-4}\ M$

    $[OH^-] = 2 \times 1.4 \times 10^{-4}\ M = 2.8 \times 10^{-4}\ M$

    $pOH = -\log(2.8 \times 10^{-4}\ M) = 3.55$

    $pH = 14 - 3.55 = 10.45$ **(Solubility equilibria)**

36. c. $pOH = pK_b + \log([NH_4^+]/[NH_3])$

    $pOH = 4.74 + \log(0.33\ M/0.44\ M) = 4.62$

    $pH = 14 - 4.62 = 9.38$ **(Buffer solutions)**

37. d. $AgNO_3(aq) + HCl(aq) \rightarrow AgCl(s) + HNO_3(aq)$

    $AgCl(s) \rightleftarrows Ag^+(aq) + Cl^-(aq)$

    $K_{sp} = 1.6 \times 10^{-10} = [Ag^+][Cl^-]$

    mol Ag^+ = 75 mL $\times$ (1 L/1000 mL) $\times$ (0.045 mol $AgNO_3$/L) $\times$ (1 mol Ag^+/1 mol $AgNO_3$) = 0.0034 mol Ag^+

    mol Cl^- = 125 mL $\times$ (1 L/1000 mL) $\times$ (1.00 mol HCl/L) $\times$ (1 mol H^+/1 mol HCl) = 0.125 mol Cl^-

    mol excess Cl^- = 0.125 mol $-$ (0.0034 mol $Ag^+ \times$ 1 mol Cl^-/1 mol Ag^+) = 0.122 mol Cl^-

    $[Cl^-]$ = 0.122 mol Cl^-/0.200 L = 0.610 M Cl^-

Compound	Initial Concentration, M	Concentration Change, M	Equilibrium Concentration, M
$[Ag^+]$	0.0	$+x$	x
$[Cl^-]$	0.610	$+x$	$0.610 + x$

    The x added to the concentration of Cl^- can be dropped because of the very small value of K_{sp} and the common ion effect.

$K_{sp} = 1.6 \times 10^{-10} = [Ag^+][Cl^-] = x \cdot 0.610\ M$

$x = 2.6 \times 10^{-10}\ M$

% $Ag_{remaining} = (2.6 \times 10^{-10}\ M/0.045\ M) \times 100 = 5.8 \times 10^{-7}$ % **(Solubility equilibrium)**

38. c. To separate the group III cations, hydrogen sulfide, H_2S, is added to a basic solution of cations. This means that both the pH and sulfide ion concentration are high. **(Qualitative analysis)**

39. b. Low pH and low sulfide ion concentration are needed to precipitate a member of qualitative analysis group II. **(Qualitative analysis)**

40. d. Mg^{2+} is a member of qual group V. **(Qualitative Analysis)**

Grade Yourself

Circle the numbers of the questions you missed, then fill in the total incorrect for each topic. If you answered more than three questions incorrectly, you need to focus on that topic. (If a topic has less than three questions and you had at least one wrong, we suggest you study that topic also. Read your textbook, a review book, or ask your teacher for help.)

Subject: Applications of Aqueous Equilibria

Topic	Question Numbers	Number Incorrect
Buffer solutions	1, 2, 3, 4, 5, 6, 7, 8, 9, 10, 11, 12, 13, 36	
Buffer capacity	14	
Titration	15, 16, 17, 18, 19, 20, 21, 22, 23, 24, 25, 26, 27	
Solubility equilibria	28, 29, 30, 31, 32, 33, 34, 35, 37	
Qualitative analysis	38, 39, 40	

Chemical Thermodynamics– Entropy, Free Energy, and Equilibria

18

 Brief Yourself

Spontaneous changes occur by themselves without outside intervention. The reverse of a spontaneous process is nonspontaneous. The driving force for spontaneous processes is a thermodynamic property called entropy. One way to define entropy, S, is a measure of the randomness or disorder in a system. Solids that have regular crystalline structures have lower entropies than liquids that only have a short-range organization of their particles, and liquids have lower entropies than gases with randomly distributed particles. Entropy is a measure of the number of possible arrangements or states that a system can take. The relationship between possible arrangements and entropy is as follows

$$S = k \ln W$$

in which S is entropy, k is Boltzman's constant, ln is the natural logarithm, and W is the number of possible arrangements of a system. This relationship is the basis for the third law of thermodynamics, which states that a perfect crystal at absolute zero has an entropy of 0 (perfect order).

The spontaneity of a chemical reaction depends on its change in entropy. The entropy change, ΔS, of a chemical reaction is the difference between the entropy of the products and the entropy of the reactants. An increase in entropy ($\Delta S > 0$) means that the products are more disorganized than the reactants. For example, if a solid decomposes to liquids and gases, the entropy of the reaction usually increases. A decrease in entropy ($\Delta S < 0$) means the products are more organized than the reactants. An increase in entropy is the driving force of spontaneity in isolated systems. Isolated systems tend spontaneously to become more random. Because chemical reactions are usually not isolated systems, the entropy changes in both the system, ΔS_{sys}, and the surroundings, ΔS_{sur}, must be considered. The second law of thermodynamics states that a spontaneous process increases the entropy of the universe, which is the sum of the entropy changes in the system and surroundings. Mathematically, this can be shown as follows.

Spontaneous Process $\Delta S_{uni} = \Delta S_{sys} + \Delta S_{sur} > 0$

The entropy of the surroundings depends on both the Kelvin temperature and enthalpy change of the system. It can be calculated as follows.

$$\Delta S_{sur} = -\Delta H/T$$

This equation shows that exothermic reactions increase the entropy of the surroundings and endothermic reactions decrease the entropy of the surroundings. At higher temperatures, the entropy change in the surroundings is smaller than those at lower temperatures. To calculate entropy changes, ΔS°, for a chemical reaction, compute the sum of the standard entropies of the products (ΣS°(products)) and subtract the sum of the standard entropies of the reactants (ΣS°(reactants)).

$$\Delta S^\circ = \Sigma S^\circ(\text{products}) - \Sigma S^\circ(\text{reactants})$$

Gibbs free energy (or free energy for short) is the most convenient thermodynamic property used to predict if reactions are spontaneous. It is defined as follows

$$G = H - TS$$

in which G is free energy, H is enthalpy, T is the Kelvin temperature, and S is entropy. The TS term represents the disorder component of energy and the enthalpy, H, represents the total energy, and what remains is the energy that can become disordered and thus can effect change and do work. Free energy is a measure of the maximum useable energy of a system. Reactions that release free energy, $\Delta G < 0$, are spontaneous and those that absorb free energy, $\Delta G > 0$, are nonspontaneous. Free energy changes are calculated from enthalpy changes, and the product of the Kelvin temperature and entropy change.

$$\Delta G^\circ = \Delta H^\circ - T\Delta S^\circ$$

Exothermic reactions in which the entropy increases are always spontaneous. Endothermic reactions in which the entropy decreases are always nonspontaneous. Reactions in which the enthalpy and entropy changes are both positive or both negative depend on the Kelvin temperature. Because ΔG° is a state function, the ΔG° of a chemical reaction is calculated from standard free energies of formation, ΔG_f^o, by using the following relationship.

$$\Delta G^\circ = \Sigma \Delta G_f^o(\text{products}) - \Sigma \Delta G_f^o(\text{reactants})$$

The relationship between free energy changes and chemical equilibrium is expressed as follows

$$\Delta G = \Delta G^\circ + RT \ln Q$$

in which ΔG is the free energy change at nonstandard conditions and Q is the reaction quotient. As a reaction proceeds it loses free energy until it reaches equilibrium. At this point ΔG equals zero; thus,

$$\Delta G^\circ = -RT \ln K$$

which states that the natural logarithm of the equilibrium constant times $-RT$ equals the standard free energy change in a reaction. When $K > 1$, the products are favored at equilibrium, ΔG° is less than zero, negative–the system releases free energy. When $K < 1$, the reactants are favored at equilibrium, ΔG° is greater than zero, positive—the system absorbs free energy.

Test Yourself

1. a. What is a spontaneous change? b. Given an example of a spontaneous change.

2. Discuss the nature of the reverse of a spontaneous change. Give an example.

3. a. If a reaction is spontaneous, does this imply that it will occur rapidly. Explain.

4. a. What is entropy? b. What type of systems have high entropies? c. Rank the three physical states in terms of their entropies.

5. a. Write an equation that can be used to calculate the absolute entropy of a substance. b. What scientist presented this equation. c. Explain the meaning the terms in the equation. d. What is the constant in the equation?

6. If a sample contains 50 diatomic molecules (near 0 K) that can only orient themselves in two possible arrangements, what is the entropy of the system?

For Problems 7 and 8, predict if the value of the entropy change, ΔS, is greater or less than zero.

7. A sample of Fe cools from 50°C to 25°C

8. HgO thermally decomposes to Hg(l) and $O_2(g)$

9. State the second law of thermodynamics in words and in an equation.

10. a. Describe the two factors that influence the entropy change that takes place in the surroundings. b. Write an equation that shows this relationship.

11. Calculate the total entropy change, ΔS_{uni}, and state if the process is spontaneous, when one mole of liquid mercury, Hg(l), changes to mercury vapor, Hg(g), at 298 K. The molar entropy of vaporization of Hg is 99 J/(K mol), and the molar enthalpy of vaporization is 59.1 kJ/mol.

12. Consider the following equation:

$$N_2(g) + 3H_2(g) \rightarrow 2NH_3(g)$$

The standard molar entropies of these substances are as follows.

Compound	S°, J/(Kmol)
$N_2(g)$	192
$H_2(g)$	131
$NH_3(g)$	193

Calculate the entropy change for this reaction.

13. a. From the following standard entropies calculate the standard entropy change, $\Delta S°$, for the complete combustion of propane, $C_3H_8(g)$, to $CO_2(g)$ and $H_2O(g)$.

$$C_3H_8(g) + 5O_2(g) \rightarrow 3CO_2(g) + 4H_2O(g)$$

Compound	S°, J/(K mol)
$C_3H_8(g)$	270
$O_2(g)$	205
$CO_2(g)$	214
$H_2O(g)$	189

b. Describe the meaning of this $\Delta S°$ value.

14. a. How is free energy defined? b. Who first presented the idea of free energy?

15. What does the change in free energy tell you about a chemical reaction?

16. Consider the following reaction:

$$H_2(g) + F_2(g) \rightarrow 2HF(g) \quad \Delta H° = -542 \text{ kJ} \quad \Delta S° = 14 \text{ J/K}$$

Calculate the $\Delta G°$ value for the reaction and state if the reaction is spontaneous.

17. Use the following thermodynamic data to calculate the $\Delta G°$ for the reaction in which $PF_5(g)$ forms from $PF_3(g)$ and $F_2(g)$ at 298 K.

$$PF_3(g) + F_2(g) \rightarrow PF_5(g)$$

Compound	$\Delta H_f°$, kJ/mol	S°, J/(K mol)
$PF_3(g)$	−919	273
$F_2(g)$	0	203
$PF_5(g)$	−1577	301

18. Consider the reaction in which solid Ag_2O decomposes to Ag(s) and $O_2(g)$.

$$Ag_2O(s) \rightarrow 2Ag(s) + \tfrac{1}{2}O_2(g) \quad \Delta H° = 31 \text{ kJ/mol}$$
$$\Delta S° = 67 \text{ J/K}$$

Over what temperature range is this reaction spontaneous?

19. Calculate the free energy change, $\Delta G°$, for the complete combustion of methane.

$$CH_4(g) + 2O_2(g) \rightarrow CO_2(g) + 2H_2O(l)$$

The standard free energies of formation for $CH_4(g)$, $CO_2(g)$, and $H_2O(l)$ are –51, –394, and –237 kJ/mol, respectively.

20. The molar entropy of fusion, ΔH_{fusion}, for water is 6.01 kJ/mol and the melting point is 273 K. Calculate the molar entropy of fusion of water.

For Problems 21 and 22, consider the following gas-phase NO_2-N_2O_4 equilibrium:

$$2NO_2(g) \rightleftarrows N_2O_4(g)$$

The $\Delta G°$ for this reaction is –6.0 kJ/mol.

21. Calculate ΔG for this equilibrium at 298 K when initially 1.0 atm of both NO_2 and N_2O_4 are present.

22. Calculate ΔG for this equilibrium at 298 K when initially 10 atm NO_2 and 5.0 atm N_2O_4 are present.

23. Consider the following gas-phase reaction:

$$H_2(g) + Cl_2(g) \rightleftarrows 2HCl(g)$$

The standard enthalpy change, $\Delta H°$, for this reaction is –92 kJ and the standard entropy change, $\Delta S°$, is –95 J/K. Calculate the equilibrium constant, K_p, at 298 K for this reaction.

24. Consider the following reaction:

$$CH_4(g) + Cl_2(g) \rightleftarrows CH_3Cl(g) + HCl(g)$$

The equilibrium constant, K_p, for this reaction is 7.58×10^{17}, and the standard enthalpy change, $\Delta H°$, is –99 kJ/mol. Calculate the entropy change, $\Delta S°$, of the reaction in J/K.

25. The enthalpy of vaporization, ΔH_{vap}, of phosphorus trichloride, $PCl_3(l)$, is 33 kJ/mol and its entropy of vaporization, ΔS_{vap}, is 95 J/(K mol). Calculate the normal boiling point of PCl_3 in °C.

26. Dinitrogen trioxide, N_2O_3, decomposes to $NO(g)$ and $NO_2(g)$ as follows.

$$N_2O_3(g) \rightarrow NO(g) + NO_2(g)$$

The standard free energy change, $\Delta G°$, for this reaction is –98 kJ, and the standard entropy change, $\Delta S°$, is 139 J/K. Calculate the temperature range over which this reaction is spontaneous.

27. The standard free energy change, $\Delta G°$, for a reaction is –39 kJ. Calculate the free energy change, ΔG, at 298 K for a Q value of 200.

28. What is the standard free energy change for a reaction that has an equilibrium constant of 0.0123 at 298 K?

29. Use the following data to calculate the K_{sp} value for AgI(s).

Species	$\Delta G_f°$, kJ/mol
AgI	–25.5
Ag^+	77
I^-	–52

30. Given the following three equations,

$$\Delta S_{uni} = \Delta S_{sur} + \Delta S_{sys}$$

$$\Delta S_{sur} = -\Delta H/T$$

$$\Delta S_{uni} = -\Delta G/T$$

derive the following equation at constant temperature.

$$\Delta G = \Delta H - T\Delta S$$

31. a. State using words the third law of thermodynamics? b. How can it be stated mathematically?

32. Consider the following change:

$$C(graphite) \rightleftarrows C(diamond)$$

The standard free energy of formation for graphite and diamond are 0.0 and 3 kJ/mol, respectively. What is the equilibrium constant for the conversion of graphite to diamond at 298 K?

33. Consider the following reaction:

    $$2N_2(g) + 5O_2(g) \rightarrow 2N_2O_5(g)$$

    Use the following data to calculate its standard entropy change. Explain the meaning of this value.

Compound	$S°$, J/(mol K)
N_2O_5	178
N_2	192
O_2	205

34. A reaction is found to have the following characteristics:

    $$\Delta S°_{sur} < 0 \qquad \Delta S°_{sys} > 0$$

    Is this reaction spontaneous or nonspontaneous? Explain.

35. Which of the following would have the lowest entropy?
    a. $H_2O(l)$
    b. $He(g)$
    c. $N_2O(g)$
    d. $NH_3(g)$
    e. all have equal entropies

36. Which of the following has the highest entropy?
    a. $H_2(g)$
    b. $F_2(g)$
    c. $Cl_2(g)$
    d. $Br_2(g)$
    e. $I_2(g)$

37. Calculate the entropy change in the surroundings, ΔS_{sur}, at 298 K, for the oxidation of carbon monoxide, $CO(g)$ to $CO_2(g)$.

    $$CO(g) + \tfrac{1}{2}O_2(g) \rightarrow CO_2(g) \qquad \Delta H° = -283 \text{ kJ}$$

    a. 950 J/(K mol)
    b. -2.83×10^5 J/(K mol)
    c. 1.04×10^3 J/(K mol)
    d. 0.949 J/(K mol)
    e. cannot be determined from this data

38. Consider the following reaction:
    $$N_2(g) + 2H_2(g) \rightarrow N_2H_4(l) \quad \Delta H° = 51 \text{ kJ}$$
    $$\Delta S° = -333 \text{ J/K}$$
    What is the free energy change for this reaction?
    a. -48.2 kJ
    b. 150 kJ
    c. -282 kJ
    d. 48.2 kJ
    e. none of these

39. Determine the temperature range over which the following reaction is spontaneous:

    $$CO(g) + \tfrac{1}{2}O_2(g) \rightarrow CO_2(g)$$

    The $\Delta H°$ for this reaction is -284 kJ and the $\Delta S°$ is -87 J/K.

    a. This reaction is spontaneous at all temperatures.
    b. This reaction is nonspontaneous at all temperatures.
    c. $T > 3.3 \times 10^3$ K
    d. $T > 3.3 \times 10^3$ K
    e. none of these

40. The enthalpy of vaporization, ΔH_{vap}, of liquid ammonia, $NH_3(l)$, is 23.3 kJ/mol and its entropy of vaporization, ΔS_{vap}, is 97.2 J/(K mol). What is the normal boiling point (1 atm) of liquid ammonia in °C?
    a. 373 K
    b. 0.240 K
    c. 2.26×10^3 K
    d. 240 K
    e. cannot be determined from this data

Check Yourself

1. a. Spontaneous changes occur by themselves without outside intervention. b. An example of a spontaneous process is that heat flows spontaneously from a hotter body to a colder one. (**Spontaneous changes**)

2. In all cases, the opposite of a spontaneous process is nonspontaneous. This means that a nonspontaneous change requires some outside driving force. For example, a body becomes warmer only after heat is added, or at 298 K, heat must be removed to change liquid water to ice. (**Spontaneous changes**)

3. Spontaneity is independent of kinetics. Some spontaneous processes occur rapidly while others proceed slowly. Reaction rates depend on activation energies, E_a. Therefore, a spontaneous process with a smaller activation energy proceeds more rapidly than one that has a larger activation energy. (**Spontaneous processes**)

4. a. Entropy is the quantitative thermodynamic property used to measure the randomness or disorder in a system. b. Systems with higher entropies are more disorganized than those with lower entropies. c. More disordered systems (less organized), such as gases, have higher values for their entropies than less disordered systems (more organized), such as solids. The structures of crystalline solids have a regular array of particles in their lattices, which means that they possess a high level of organization. (**Entropy**)

5. a and b. $S = k \ln W$ is the Boltzman equation. c. S is the entropy of a substance, k is Boltzman's constant (1.381×10^{-23} J/K), ln is the natural log, and W is number of possible arrangements or states (sometimes called microstates) for a system. d. The Boltzman constant is the ratio of the ideal gas constant and Avogadro's number. (**Entropy**)

6. The value of W is 2^{50} ($2 \times 2 \times 2 \times 2 \ldots$) or 1.1×10^{15}.
   $S = k \ln W = 1.38 \times 10^{-23}$ J/K ln (1.1×10^{15}) = 4.8×10^{-22} J/K (**Entropy**)

7. As Fe atoms cool, their average kinetic energy decreases, which means a smaller number of possible arrangements exists. In other words, a system at a lower thermal energy is less random than one at a higher thermal energy. Thus, ΔS is less than zero (negative). (**Entropy changes**)

8. In decomposition reactions, one reactant breaks down to two or more products. In this case solid HgO changes to liquid Hg and gaseous O_2.

   $$HgO(s) \xrightarrow{\Delta} Hg(l) + \tfrac{1}{2}O_2(g)$$

   Both products are more disorganized than HgO because liquids and gases are more disordered that crystalline solids. Thus, ΔS is greater than zero (positive). (**Entropy changes**)

9. All spontaneous processes increase the entropy of the universe. Mathematically, this can be shown as $\Delta S_{uni} = \Delta S_{sys} + \Delta S_{sur} > 0$ for any spontaneous process. (**Entropy changes**)

10. a. One factor is the maximum heat that can be transferred from the system to the surroundings at constant pressure, q_P. If the system loses heat to the surroundings, the additional thermal energy disorganizes the surroundings and if the system gains heat, the surroundings lose heat causing less disorganization. Thus, ΔS_{sur} is proportional to $-q_P$. The second factor is the Kelvin temperature, T. If the surroundings is at a high temperature, then it is highly disorganized already from the thermal energy and the heat flow from the system has little effect on the entropy change.

    b. $\Delta S_{sur} = -\Delta H_{sys}/T$ (**Entropy changes**)

11. $\Delta S_{sur} = -H_{sys}/T$
    $\Delta S_{sur} = -59{,}100$ J/mol/298 K = -198 J/(K mol)

ΔS_{uni} = 99 J/(K mol) + (–198 J/(K mol)) = –99 J/(K mol)

The answer, –99 J/(K mol), shows that the vaporization of Hg(l) at 298 K is nonspontaneous. (**Second law**)

12. $\Delta S^o = \Sigma S^o$(products) – ΣS^o(reactants)

    ΔS^o = [2 mol × $S^o(NH_3)$] – [(1 mol × $S^o(N_2)$) + (3 mol × $S^o(H_2)$)]

    ΔS^o = [2 mol × 193 J/(K mol)] – [(1 mol × 192 J/(K mol)) + (3 mol × 131 J/(K mol))] = –199 J/(K mol) (**Entropy changes**)

13. $\Delta S^o = \Sigma S^o$(products) – ΣS^o(reactants)

    ΔS^o = [(3 mol × $S^o(CO_2)$) + (4 mol × $S^o(H_2O)$)] – [(1 mol × $S^o(C_3H_8)$) + (5 mol × $S^o(O_2)$)]

    ΔS^o = [(3 mol × 214 J/(K mol)) + (4 mol × 189 J/(K mol))] – [(1 mol × 270 J/(K mol)) + (5 mol × 205 J/(K mol))] = 103 J/(K mol)

    b. Because the entropy change is positive, the system goes from a less random to a more random state. In part, this may be attributed to the production of seven moles of gaseous products from six moles of gaseous reactants. (**Entropy changes**)

14. a. Free energy is defined as: $G = H - TS$. In this equation, G is free energy, H is enthalpy, T in the Kelvin temperature, and S is entropy. b. J. W. Gibbs first proposed the idea of free energy. (**Free energy**)

15. The change in free energy is the maximum useable energy that a reaction can transfer. In other words, it is the maximum work, w_{max}, that a system can do on the surroundings.

    $\Delta G = w_{max}$ (**Free energy**)

16. $\Delta G^o = \Delta H^o - T\Delta S^o$ = –542 kJ – (298 K × 14 J/K × 1 kJ/1000 J) = –546 kJ

    The answer, –546 kJ, is less than zero; hence, this reaction is spontaneous. (**Gibbs equation**)

17. $\Delta H^o = \Sigma \Delta H_f^o$(products) – $\Sigma \Delta H_f^o$ (reactants)

    ΔH^o = (1 mol × $\Delta H_f^o(PF_5)$) – [(1 mol × $\Delta H_f^o(PF_3)$) + (1 mol × $\Delta H_f^o(F_2)$)]

    ΔH^o = –1577 kJ – (–919 kJ + 0 kJ) = –658 kJ

    $\Delta S^o = \Sigma S^o$(products) – ΣS^o(reactants)

    ΔS^o = (1 mol × $S^o(PF_5)$) – [(1 mol × $S^o(PF_3)$) + (1 mol × $S^o(F_2)$)]

    ΔS^o = (1 mol × 301 J/(K mol)) – [(1 mol × 273 J/(K mol)) + (1 mol × 203 J/(K mol))] = –175 J/K

    $\Delta G^o = \Delta H^o - T\Delta S^o$ = –658 kJ – (298 K × –0.175 J/K × 1 kJ/1000 J) = –606 kJ (**Gibbs equation**)

18. $\Delta G = \Delta H^o - T\Delta S^o$

    0 = 31 kJ – (T × 67 J/K × 1 kJ/1000 J)

    $T = 4.6 \times 10^2$ K

    For temperatures above 4.6×10^2 K, the decomposition of silver oxide is spontaneous, and for temperatures below that, it is nonspontaneous. (**Gibbs equation**)

19. $\Delta G^o = \Sigma \Delta G_f^o$(products) – $\Sigma \Delta G_f^o$(reactants)

    ΔG^o = [(1 mol × $\Delta G_f^o(CO_2)$) + (2 mol × $\Delta G_f^o(H_2O(l))$)] – [(1 mol × ΔG_f^o (CH_4)) × (2 mol $\Delta G_f^o(O_2)$)]

    ΔG^o = [(1 mol × –394 kJ/mol) + (2 mol × –237 kJ/mol)] – [(1 mol × –51 kJ/mol) + 2 mol × 0 kJ/mol][1] – 817 kJ (**Free energy changes**)

20. $\Delta S_{fus} = \Delta H_{fus}/T_{mp}$ = 6.01 kJ/mol/273 K = 0.0220 kJ/(K mol) = 22.0 J/(K mol) (**Entropy changes**)

21. $Q = P_{N_2O_4}/P_{NO_2}$ = 1.0 atm/(1.0 atm)^2 = 1.0

    $\Delta G = \Delta G^o + RT \ln Q$ = –6.0 kJ/mol + $RT \ln 1$ = –6.0 kJ/mol (**Nonstandard free energy changes**)

22. $Q = P_{N_2O_4}/P_{NO_2} = 5.0$ atm/$(10$ atm$)^2 = 0.050$

    $\Delta G = \Delta G^o + RT \ln Q$

    $\Delta G = -6.0$ kJ/mol $+ (8.314$ J/(mol K) $\times$ (1 kJ/1000 J) $\times$ 298 K $\times \ln 0.050)$

    $\Delta G = -6.0$ kJ/mol $+ (-7.4$ kJ/mol$) = -13.4$ kJ/mol   (**Nonstandard free energy changes**)

23. $\Delta G^o = \Delta H^o - T\Delta S^o$

    $\Delta G^o = -92$ kJ $- (298$ K $\times -95$ J/K $\times$ 1 kJ/1000 J$) = -64$ kJ

    $\Delta G^o = -RT \ln K_p$

    -64 kJ $= -8.314$ J/K $\times$ (1 kJ/1000 J) $\times 298$ $K \times \ln K_p$

    $\ln K_p = 26$

    $K_p = 1.7 \times 10^{11}$ (ΔG^o and K) (**Free energy change - equilibrium relationship**)

24. $\Delta G^o = -RT \ln K_p$

    $\Delta G^o = -8.314$ J/(mol K) $\times$ (1 kJ/1000 J) $\times$ 298 K $\times \ln 7.58 \times 10^{17} = -102$ kJ

    $\Delta G^o = \Delta H^o - T\Delta S^o$

    -102 kJ $= -99$ kJ $- (298$ K $\times S^o)$

    $\Delta S^o = 0.0010$ kJ/K $= 1.0$ J/K (ΔG^o and K) (**Free energy change-equilibrium relationship**)

25. $\Delta S_{vap} = \Delta H_{vap}/T_{bp}$

    $T_{bp} = \Delta H_{vap}/\Delta S_{vap} = 33$ kJ/mol/(95 J/(K mol) $\times$ 1 kJ/1000 J $= 347$ K $= 74°$C (**Entropy changes**)

26. $\Delta G° = \Delta H^o - T\Delta S^o$

    -98 kJ $= \Delta H^o - (298$ K $\times 139$ J/K $\times$ 1 kJ/1000 J)

    $\Delta H^o = -56.6$ kJ

    An exothermic reaction with a positive entropy change is spontaneous at all temperatures. (**Gibbs equation**)

27. $\Delta G = \Delta G^o + RT \ln Q$

    $\Delta G = -39$ kJ $+ (8.314$ J/(mol K) $\times$ 1 kJ/1000 J $\times$ 298 K $\times \ln 200)$

    $\Delta G = -39$ kJ $+ (13.1$ kJ$) = -25.9$ kJ (**Nonstandard free energy changes**)

28. $\Delta G^o = -RT \ln K_p$

    $\Delta G^o = -8.314$ J/(mol K) $\times$ (1 kJ/1000 J) $\times$ 298 K $\times \ln 0.0123 = 10.9$ kJ (**Free energy-equilibrium relationship**)

29. $\Delta G^o = (77$ kJ $+ (-52$ kJ$)) - (-25.5$ kJ$) = 50.5$ kJ

    $\Delta G^o = -RT \ln K_{sp}$

    50.5 kJ $= -8.314$ J/(mol K) $\times$ 1 kJ/1000 J $\times$ 298 K $\times \ln K_{sp}$

    $K_{sp} = 4.1 \times 10^{-21}$ (**Free energy-equilibrium relationship**)

30. $\Delta S_{uni} = \Delta S_{sur} + \Delta S_{sys}$

    $\Delta S_{sur} = -\Delta H/T$

    $\Delta S_{uni} = -\Delta G/T$

    $-\Delta G/T = -\Delta H/T + \Delta S_{sys}$

    $-T(-\Delta G/T) = -T(-\Delta H/T + \Delta S_{sys})$

    $\Delta G = \Delta H - T\Delta S$ (**Gibbs relationship**)

31. a. The entropy of a perfect crystal at absolute zero is 0 J/K. b. $S = k \ln W$, where S is the absolute entropy, k is the Boltzman constant, and W is the number of microstates for a system. (**Third law of thermodynamics**)

32. $\Delta G° = 3$ kJ $- 0$ kJ $= 3$ kJ

    $\Delta G° = -RT \ln K_p$

    3 kJ $= -8.314$ J/(mol K) $\times$ (1 kJ/1000 J) $\times$ 298 K $\times \ln K$

    $K = 0.062$ (**Free energy-equilibrium relationship**)

33. $\Delta S° = \Sigma S°(\text{products}) - \Sigma S°(\text{reactants})$

    $\Delta S° = (2 \times 178$ J/(K mol)) $- ((2$ mol $\times 192$ J/(K mol)) $+ (5$ mol $\times 205$ J/(K mol))) $= -1053$ J/(K mol)

    The high negative value for the entropy change is the result of going from seven moles of reactant particles to only two moles of products. (**Entropy changes**)

34. $\Delta S°_{sur} < 0 \quad \Delta S°_{sys} > 0$

    For a reaction to be spontaneous, the sum of the entropy changes in the system and surroundings must be greater than zero ($\Delta S_{uni} = \Delta S_{sys} + \Delta S_{sur} > 0$). If the magnitude of the entropy change in the surroundings is a greater negative value than the positive value for the entropy change in system, the reaction is nonspontaneous. If the reverse is true, the reaction is spontaneous; thus from this data, there is no way to know if the reaction is spontaneous. (**Free energy changes**)

35. a. $H_2O(l)$ has the lowest entropy because it is a liquid; thus, it is more organized. (**Entropy**)

36. e. $I_2(g)$ has the highest entropy because it has the largest number of electrons. (**Entropy**)

37. a. $\Delta S_{sur} = -\Delta H_{sys}/T$

    $\Delta S_{sur} = -(-283,000$ J/mol$)/298$ K $= 950$ J/(K mol) (**Entropy change**)

38. b. $\Delta G° = \Delta H° - T\Delta S° = 51$ kJ $- (298$ K $\times -333$ J/K $\times 1$ kJ/1000 J) $= 150$ kJ

    Because $\Delta G°$ is a positive value, this reaction is nonspontaneous at standard conditions. (**Free energy change**)

39. d. $T < 3.3 \times 10^3$ K

    $\Delta G = \Delta H° - T\Delta S°$

    $0 = -284$ kJ $- (T \times -87$ J/K $\times 1$ kJ/1000 J)

    $T = 3.3 \times 10^3$ K

    At temperatures below 3.3×10^3 K, the $-T\Delta S°$ term is less positive than the $\Delta H°$ term is negative. Hence, the value of ΔG is less than zero which makes it spontaneous. (**Gibbs equation**)

40. d. $\Delta S_{vap} = \Delta H_{vap}/T_{bp}$

    $T_{bp} = \Delta H_{vap}/\Delta S_{vap} = 23.3$ kJ/mol$/(97.2$ J/(K mol)) $\times 1$ kJ/1000 J $= 240$ K (**Entropy changes**)

Grade Yourself

Circle the numbers of the questions you missed, then fill in the total incorrect for each topic. If you answered more than three questions incorrectly, you need to focus on that topic. (If a topic has less than three questions and you had at least one wrong, we suggest you study that topic also. Read your textbook, a review book, or ask your teacher for help.)

Subject: Chemistry Fundamentals

Topic	Question Numbers	Number Incorrect
Spontaneous changes	1, 2, 3	
Entropy	4, 5, 6, 35, 36	
Entropy changes	7, 8, 9, 10, 12, 13, 20, 25, 33, 37, 40	
Second law	11	
Free energy	14, 15	
Free energy changes	19, 34, 38	
Free energy-equilibrium relationship	23, 24, 28, 29, 32	
Gibbs equation	16, 17, 18, 26, 39	
Gibbs relationship	30	
Nonstandard free energy changes	21, 22, 23, 24, 27	
Third law of thermodynamics	31	

Electrochemistry

Brief Yourself

Electrochemistry is the study of both the conversion of chemical energy to electrical energy in voltaic cells (galvanic cells) and the conversion of electrical energy to chemical energy in electrolytic cells.

A simple voltaic cell consists of two electrodes immersed in electrolyte solutions connected by an external circuit and a salt bridge or a porous cup. One electrode is the anode, the site of oxidation, and the other is the cathode, the site of reduction. The anode is the negative electrode because it is the source of electrons. The cathode is the positive electrode because it takes in electrons. The electrons from the anode flow through an external circuit and are taken in by the cathode. At the cathode the electrons are accepted by the cations in solution. The salt bridge, either a tube filled with inert salt such as $NaNO_3$ or a porous barrier, completes the circuit. Cations migrate towards the cathode and anions move towards the anode. This flow of ions completes the electric current and maintains the neutrality of the cell.

The capacity of a voltaic cell to push electrons through a circuit is called the electromotive force (emf) of the cell or the cell potential, ξ. The SI unit of electromotive force is volts, V. A potentiometer is used to measure accurately the voltage of a voltaic cell. The cell potential may be calculated from standard reduction potentials, ξ^o. A standard reduction potential is the voltage produced by a reduction half-cell at standard conditions relative to the standard hydrogen electrode, which is assigned a value of 0.0 V. The half-reaction for the standard hydrogen electrode is as follows.

$$2H^+(aq, 1\ M) + 2e^- \rightarrow H_2(g,\ 1\ atm)\quad \xi^o_{red} = 0.0\ V$$

The more positive the standard reduction potential, the more readily a substance undergoes reduction–a better oxidizing agent. The more negative the standard reduction potential, the more readily a substance undergoes oxidation–a better reducing agent. A cell potential is calculated by subtracting the standard reduction potential of the anode from that of the cathode.

$$\xi^o_{cell} = \xi^o (cathode) - \xi^o (anode)$$

All voltaic cells are spontaneous; thus, the standard cell potential, ξ^o_{cell}, always has a positive value.

Reactions with positive ξ values are spontaneous, those with negative ξ values are nonspontaneous, and those that equal zero are at equilibrium. The relationship between ΔG^o and ξ^o is as follows

$$\Delta G^o = -nF\xi^o$$

in which $\xi°$ is the standard cell potential, $\Delta G°$ is the standard free energy change, n is the number of moles of electrons transferred, and F is Faraday's constant (96,500 C/mol e⁻). In the nonstandard state, this equation is as follows.

$$\Delta G = -nF\xi$$

Because $\xi°$ is related to $\Delta G°$, and $\Delta G°$ is related to the equilibrium constant, then $\xi°$ is related to K as follows.

$$\Delta G° = \frac{-RT}{nF} \ln K$$

The more positive the $\xi°$ value the farther the equilibrium lies towards pure products, and the more negative the $\xi°$ value the closer the equilibrium lies towards pure reactants. At all other conditions except for standard conditions, the Nernst equation is used.

$$\xi = \xi° - \frac{RT}{nF} \ln Q$$

In the Nernst equation, ξ is the cell potential, $\xi°$ is the standard cell potential, and Q is the reaction quotient.

In an electrolytic cell, electrical energy changes to chemical energy. All electrolytic cells are non-spontaneous, $\Delta G° > 0$, and require an outside energy source. Electricity is pumped into the cathode, the site of reduction, and pumped out of the anode, the site of oxidation. In an electrolytic cell, the substance that most readily reduces plates the cathode and the substance that most readily oxidizes forms at the anode. For example, when molten sodium chloride, NaCl(l), undergoes electrolysis Na plates the cathode and $Cl_2(g)$ bubbles up from the anode. If an electrolytic cell contains water $O_2(g)$ is produced at the anode and $H_2(g)$ is given off at the cathode. Electrolysis of aqueous NaCl, or brine, results in the production of $Cl_2(g)$ at the anode and $H_2(g)$ at the cathode. The amounts of products produced at the electrodes is predicted using Faraday's law, which states that the number of moles of products formed during electrolysis is directly proportional to the number of moles of electrons that pass through the cell.

Test Yourself

1. Describe the major difference between voltaic (galvanic) cells and electrolytic cells.

2. a. What process takes place at the anode of a voltaic cell? b. What process takes place at the cathode of a voltaic cell?

3. Write the cell notation for a Zn-Cu voltaic cell in which Zn is immersed in a Zn^{2+} solution at the anode, and Cu is immersed in a Cu^{2+} solution at the cathode.

4. Consider the following cell notation:

   Al(s)|Al^{3+}(aq)||Ni^{2+}(aq)|Ni(s)

   a. What substances are at the anode? b. What substances are at the cathode? c. Write the net ionic equation for the cell reaction.

5. A voltaic cell is constructed with Pb(s) and Pb(NO₃)₂(aq) for one half-cell and Zn(s) and Zn(NO₃)₂(aq) for the other. If Zn is a better reducing agent than Pb and the salt bridge contains an aqueous solution of KNO₃, answer each of the following.

   a. Which substance is the anode?

b. Which substance is the cathode?

c. Which ions move toward the anode?

d. Which ions move toward the cathode?

e. Which metal is the negative electrode?

f. Which metal is the positive electrode?

g. Write the cell notation for the reaction.

6. a. What is a volt?

   b. How are joules calculated from the voltage?

7. a. Describe the standard half-cell that is used in electrochemistry to measure standard potentials.
   b. Write an equation for its half reaction and show its standard potential.

8. a. If the standard reduction potentials for Zn and Cu are –0.76 V and 0.34 V, respectively, what is the standard cell potential? b. Write an equation for the cell reaction.

9. The standard reduction potentials of Al and Ni are –1.66 and –0.28 V, respectively. Is Al a stronger or weaker reducing agent than Ni? Explain.

10. The standard reduction potentials for Zn^{2+}, Mg^{2+}, and Na^+ are –0.76, –2.37, and –2.71 V, respectively. Which of the following is the strongest oxidizing agent: Zn^{2+}, Mg^{2+}, or Na^+?

11. The standard reduction potentials for Al and Pb are –1.66 and –0.13 V, respectively. Calculate the cell potential for a voltaic cell that has Al and Pb electrodes and show the half reactions and overall cell reaction.

For Problems 12 to 15 consider a voltaic cell constructed with the following substances:

$Cr^{3+}(aq) + 3e^- \rightarrow Cr(s)$ $\xi° = -0.74$ V

$MnO_4^-(aq) + 8H^+(aq) + 5e^- \rightarrow Mn^{2+}(aq) + 4H_2O(l)$
$\xi° = +1.51$ V

12. Which substances are oxidized and reduced in this cell?

13. Write the overall cell equation and calculate the cell potential.

14. Which are the negative and positive electrodes?

15. Write the cell notation for this voltaic cell.

16. The standard reduction potential for Cu^{2+} is +0.32 V. If the standard reduction potential of Ni^{2+} is –0.28 V, will a strip of Cu dissolve in a Ni^{2+} solution?

17. a. What is the relationship between the free energy change and standard cell potential?
    b. Describe each term in the equation.

18. If the $\xi°$ value for a redox reaction is +1.00 V and 2.00 moles of electrons are transferred in the reaction. What is the free energy change for the reaction?

19. The standard reduction potentials for Cl_2 and I_2 are as follows:

$Cl_2(g) + 2e^- \rightarrow 2Cl^-(aq)$ $\xi° = +1.36$ V
$I_2(g) + 2e^- \rightarrow 2I^-(aq)$ $\xi° = +0.54$ V

Calculate the standard free energy change, $\Delta G°$, for the following reaction and determine if the reaction is spontaneous.

$Cl_2(g) + 2I^-(aq) \rightarrow I_2(s) + 2Cl^-(aq)$

20. The anode and cathode half reactions for the lead-storage battery are as follows:

$Pb(s) + SO_4^{2-}(aq) \rightarrow PbSO_4(s) + 2e^-$
$\xi° = +0.36$ V

$PbO_2(s) + SO_4^{2-}(aq) + 4H^+(aq) + 2e^- \rightarrow PbSO_4(s) + 2H_2O(l)$
$\xi° = +1.68$ V

Calculate the equilibrium constant, K, for the lead storage battery at 298 K.

21. A voltaic cell has Sn and Ag electrodes. Predict the voltage of this cell at 298 K when $[Sn^{2+}] = 0.0010$ M and $[Ag^+] = 0.75$ M. The standard reduction potentials for Ag^+ and Sn^{2+} are 0.80 V and -0.14 V, respectively.

22. The cathode of a voltaic cell has a strip of Fe(s) immersed in an unknown concentration of $Fe^{3+}(aq)$ and the anode is a standard Zn half-cell. The potential of the cell is 0.65 V. If the standard cell potential for a Zn/Fe cell is 0.72 V, what is the molar concentration of Fe^{3+}?

23. A voltaic cell has 1.0 M Zn^{2+} at the cathode and 0.15 M Zn^{2+} at the anode. If the standard reduction potential for Zn is -0.76 V, what is the cell potential?

24. What is electrolysis? What is an electrolytic cell?

25. a. Write the anode reaction when liquid NaCl undergoes electrolysis. b. Write the cathode reaction when liquid NaCl undergoes electrolysis. c. What is the overall cell reaction in the electrolysis of liquid NaCl?

26. Write two possible half reactions that occur at the cathode when water with added acid undergoes electrolysis.

27. A solution of copper(II) sulfate, $CuSO_4(aq)$, undergoes electrolysis using inert Pt electrodes. On one electrode O_2 bubbles evolve and Cu plates out on the other. Explain why these products result, and write the anode, cathode, and overall balanced equation.

28. A solution of sodium sulfate, $Na_2SO_4(aq)$, undergoes electrolysis using inert Pt electrodes. O_2 bubbles evolve from one electrode and H_2 bubbles are released from the other. Write the anode, cathode, and overall balanced equation.

29. a. What is the unit of electric current? b. How is the number of coulombs of charge found? c. What is the meaning of the Faraday constant?

30. a. Write the net ionic equation for the reaction of Pb with Zn^{2+}. b. What is the standard potential for this reaction? c. Will a strip of Pb(s) dissolves in a 1 M Zn^{2+} solution?

$$Pb^{2+}(aq) + 2e^- \rightarrow Pb(s) \quad \xi° = -0.13 \text{ V}$$

$$Zn^{2+}(aq) + 2e^- \rightarrow Zn(s) \quad \xi° = -0.76 \text{ V}$$

31. How many coulombs and moles of electrons pass through an electrolytic cell when 10.0 A passes through the cell for 1.00 min?

32. What mass of Pb forms at the cathode when 5.00 A flow through a solution of $Pb(NO_3)_2$ for 25.0 min?

33. How many minutes does it take to produce 35.0 g Fe by passing a 20.0 A current through a solution of $Fe(NO_3)_3(aq)$?

34. a. What are fuel cells? b. Give an example of a common fuel cell.

For Problems 35 to 38, use the following information: A voltaic cell is constructed using Mg(s) and $Mg(NO_3)_2(aq)$ for one half-cell and Cd(s) and $Cd(NO_3)_2(aq)$ for the other. Mg is a better reducing agent than Cd and a salt bridge connects the half reaction.

35. Which substance is the anode?
    a. Mg
    b. Cd
    c. NO_3^-
    d. $NaNO_3$
    e. none of these

36. Which ions move toward the cathode?
    a. Mg^{2+} only
    b. Mg^{2+}, Cd^{2+}
    c. Cd^{2+} only
    d. NO_3^- only
    e. none of these

37. Which is the negative electrode?
   a. Mg
   b. Cd
   c. NO$_3^-$
   d. NaNO$_3$
   e. none of these

38. What is its cell notation?
   a. Cd(s)|Cd^{2+}(aq)||Mg^{2+}(aq)|Mg(s)
   b. Cd(s)|Mg^{2+}(aq)||Cd^{2+}(aq)|Mg^{2+}(s)
   c. Mg(s)|Mg^{2+}(aq)||Cd^{2+}(aq)|Cd(s)
   d. Mg(s)|Cd^{2+}(aq)||Mg^{2+}(aq)|Cd(s)
   e. none of these

39. Which of the following substances has the highest standard reduction potential?
   a. Li
   b. Cu
   c. Br$_2$
   d. F$_2$
   e. H$_2$

40. Consider the following half reactions:

   Fe(OH)$_2$(s) + 2e$^-$ → Fe(s) + 2OH$^-$(aq) $\xi°= -0.88$ V

   NiO$_2$(s) + 2H$_2$O(l) + 2e$^-$ → Ni(OH)$_2$(s) + 2OH$^-$(aq) $\xi° = +0.49$ V

   What is the standard cell potential for a voltaic cell using these electrodes?
   a. 0.49 V
   b. –0.39 V
   c. 1.37 V
   d. –1.37 V
   e. none of these

Check Yourself

1. In a voltaic cells, also called galvanic cells, chemical energy is converted to electrical energy, and in electrolytic cells electrical energy is converted to chemical energy. (**Electrolytic cells**)

2. Oxidation, the loss of electrons, occurs at the anode, and reduction, gaining electrons, occurs at the cathode. (**Voltaic cells**)

3. Zn(s)|Zn^{2+}(aq)||Cu^{2+}(aq)|Cu(s) (**Cell notation**)

4. a. Al oxidizes to Al^{3+} at the anode.
   b. Ni^{2+} reduces to Ni at the cathode
   c. 2Al(s) + 3Ni^{2+}(aq) → 3Ni(s) + 2Al^{3+}(aq) (**Cell notation**)

5. a. The anode is the site of oxidation; hence, Zn(s) is oxidized to Zn^{2+}(aq) at the anode.
   b. The cathode is the site of reduction; hence, Pb^{2+}(aq) is reduced to Pb(s) on the cathode.
   c. Anions migrate towards the anode. Therefore, the NO$_3^-$ ions from the salt bridge and from the Pb(NO$_3$)$_2$ solution move toward the anode.
   d. Cations migrate toward the cathode. Thus, the Zn^{2+} produced from the oxidation of Zn, and K$^+$ ions from the salt bridge move towards the cathode.
   e. The anode is the negative electrode; thus, Zn is the negative electrode.
   f. The cathode is the positive electrode; therefore, Pb is the positive electrode.
   g. Zn(s)|Zn^{2+}(aq)||Pb^{2+}(aq)|Pb(s) (**Voltaic cells**)

6. a. 1 V = 1 J/1 C One volt is the emf just needed to give 1 J of energy to that number of electrons having a charge of 1 coulomb, C–the charge on 6.2×10^{18} electrons. b. J = V × C **(Volts)**

7. a. The standard half-cell is the standard hydrogen electrode, SHE, in which H^+ ions are reduced to hydrogen gas, $H_2(g)$. b. $H_2(g) \rightarrow 2H^+(aq) + 2e^-$ $\xi° = 0.0$ V **(Cell potentials)**

8. a. $\xi° = +0.34$ V $-$ (-0.76 V) $= +1.10$ V
   b. $Zn(s) + Cu^{2+}(aq) \rightarrow Cu(s) + Zn^{2+}(aq)$ **(Cell potentials)**

9. A reducing agent releases electrons. Thus, a stronger reducing agent more readily undergoes oxidation than a weaker one. Because Al has a more negative standard reduction potential than Ni, then Al is a stronger reducing agent than Ni. **(Reduction potentials)**

10. The strongest oxidizing agent is the one that most readily undergoes reduction–the one with the highest $\xi°$ value. Because Zn^{2+} has the least negative $\xi°$ value it is the strongest oxidizing agent in this group. **(Reduction potentials)**

11. Because the standard reduction potential of Pb^{2+} is less negative (more positive) than that of Al^{3+}, Pb^{2+} undergoes reduction and Al undergoes oxidation. Thus, the Al/Al^{3+} half reaction must be reversed and the sign of its $\xi°$ must be changed.

    $2Al(s) \rightarrow 2Al^{3+}(aq) + 6e^-$ $\quad\quad\quad\quad\quad \xi° = +1.66$ V

    $3Pb^{2+}(aq) + 6e^- \rightarrow 3Pb(s)$ $\quad\quad\quad\quad\quad \xi° = -0.13$ V

    $2Al(s) + 3Pb^{2+}(aq) \rightarrow 3Pb(s) + 2Al^{3+}(aq)$ $\xi°_{cell} = +1.53$ V   **(Cell potentials)**

12. The large positive value of the MnO_4^-/Mn^{2+} half reaction indicates that $MnO_4^-(aq)$ undergoes reduction. Hence, Cr undergoes oxidation because of its negative value. When the Cr/Cr^{3+} half reaction is reversed, the sign of its half-cell potential is changed.

    $Cr(s) \rightarrow Cr^{3+}(aq) + 3e^-$ $\xi° = +0.74$ V

    $MnO_4^-(aq) + 8H^+(aq) + 5e^- \rightarrow Mn^{2+}(aq) + 4H_2O(l)$ $\xi° = +1.51$ V   **(Reduction potentials)**

13. To obtain the overall cell reaction, equalize the electrons and add the half reactions. To obtain the standard cell potential, add the half-cell potentials.

    $5Cr(s) \rightarrow 5Cr^{3+}(aq) + 15e^-$ $\quad\quad\quad\quad\quad\quad\quad\quad\quad\quad\quad\quad\quad\quad \xi° = +0.74$ V

    $3MnO_4^-(aq) + 24H^+(aq) + 15e^- \rightarrow 3Mn^{2+}(aq) + 12H_2O(l)$ $\quad\quad \xi° = +1.51$ V

    $5Cr(s) + 3MnO_4^-(aq) + 24H^+(aq) \rightarrow 3Mn^{2+}(aq) + 5Cr^{3+}(aq) + 12H_2O(l)$ $\quad \xi° = +2.25$ V

    (Cell reaction and potential)

14. Chromium, Cr, is oxidized to Cr^{3+}, thus it is the anode–the negative electrode. The cathode, the positive electrode, is an inert metal such as Pt immersed in a solution of acidic permanganate. An inert metal is needed to provide the solid surface for the electron transfer. At the Pt cathode, the purple MnO_4^- ions take in electrons, producing the almost colorless Mn^{2+} ions. (Anode and cathode)

15. The cell notation is $Cr(s)|Cr^{3+}(aq)||MnO_4^-(aq), Mn^{2+}(aq)|Pt$. **(Cell notations)**

16. $\xi° = -0.28$ V $+ (-0.32$ V$) = -0.60$ V Because the standard potential is negative Cu does not dissolve in a Ni^{2+} solution. **(Standard potentials)**

17. a. $\Delta G° = -nF\xi°$ b. $\Delta G°$ is the standard free energy change, n is the number of electrons transferred in a reaction, F is Faraday's constant, and $\xi°$ is the standard cell potential. **(Cell potential and thermodynamics)**

18. $\Delta G° = -nF\xi°$

$\Delta G° = -(2.00 \text{ mol e}^- \times 9.65 \times 10^4 \text{ J/(V mol e}^-) \times 1.00 \text{ V}) = -1.93 \times 10^5 \text{ J} \times (1 \text{ kJ}/1000 \text{ J}) = -193 \text{ kJ}$

(Cell potential and thermodynamics)

19. $\xi° = \xi°(Cl_2) + \xi°(I_2) = +1.36 \text{ V} + (-0.54 \text{ V}) = 0.82 \text{ V}$

Two moles of electrons ($n = 2$ mol) are transferred in the reaction.

$\Delta G° = -nF\xi° = -(2 \text{ mol e}^- \times 96{,}500 \text{ J/(C mol e}^-) \times 0.82 \text{ V}) = -1.6 \times 10^5 \text{ J} = -1.6 \times 10^2 \text{ kJ}$

The answer, a $\Delta G° < 0$, shows that this reaction is spontaneous at standard conditions. **(Cell potential and thermodynamics)**

20. $Pb(s) + SO_4^{2-}(aq) \rightarrow PbSO_4(s) + 2e^-$ $\quad\quad\quad\quad\quad\quad\quad\quad\quad\quad\quad\quad\quad \xi° = +0.36 \text{ V}$

$PbO_2(s) + SO_4^{2-}(aq) + 4H^+(aq) + 2e^- \rightarrow PbSO_4(s) + 2H_2O(l)$ $\quad\quad \xi° = +1.68 \text{ V}$

$Pb(s) + PbO_2(s) + 4H^+(aq) + SO_4^{2-}(aq) \rightarrow 2PbSO_4(s) + 2H_2O(l)$ $\quad \xi° = +2.04 \text{ V}$

$\xi° = (0.02569/n) \times \ln K$

$+2.04 \text{ V} = (0.02569/2) \times \ln K$

$\ln K = 159$

antiln ($\ln K$) = antiln (159)

$K = e^{159} = 1.13 \times 10^{69}$ **(Standard potential and equilibrium constant relationship)**

21. The overall cell reaction.

$$Sn(s) + 2Ag^+(aq) \rightarrow 2Ag(s) + Sn^{2+}(aq)$$

$\xi°_{cell} = \xi°(\text{cathode}) - \xi°(\text{anode}) = +0.80 \text{ V} - (-0.14 \text{ V}) = +0.94 \text{ V}$

$\xi = \xi° - 0.02569/n \times \ln ([Sn^{2+}]/[Ag^+]^2)$

$\xi = +0.94 \text{ V} - ((0.02569/2 \text{ mol e}^-) \ln [0.0010 \text{ M}/(0.75 \text{ M})^2])$

$\xi = +0.94 \text{ V} - (-0.081 \text{ V}) = +1.02 \text{ V}$ **(Nernst equation)**

22. $Zn(s) \rightarrow Zn^{2+}(aq) + 2e^-$

$Fe^{3+}(aq) + 3e^- \rightarrow Fe(s)$

Equalizing the electrons that are released and taken in shows that six electrons transfer from the anode to the cathode ($n = 6$). Adding the two half reactions gives the following overall cell equation.

$$3Zn(s) + 2Fe^{3+}(aq) \rightarrow 2Fe(s) + 3Zn^{2+}(aq)$$

$\xi = \xi° - 0.02569/n \times \ln ([Zn^{2+}]^3/[Fe^{3+}]^2)$

$0.65 \text{ V} = +0.72 \text{ V} - 0.02569/6 \text{ mol e}^- \ln (1.0 \text{ M})^3/[Fe^{3+}]^2$

$-0.07 \text{ V} = -0.02569/6 \text{ mol e}^- \ln (1.0 \text{ M})^3/[Fe^{3+}]^2$

$\ln (1/[Fe^{3+}]^2) = 16.4$

$[Fe^{3+}] = 3 \times 10^{-4} \text{ M}$ **(Nernst equation)**

23. (Anode) $Zn(s) \rightarrow Zn^{2+}(aq) + 2e^-$ $\quad\quad\quad\quad\quad\quad \xi° = +0.76 \text{ V}$

(Cathode) $Zn^{2+}(aq) + 2e^- \rightarrow Zn(s)$ $\quad\quad\quad\quad\quad\quad \xi° = -0.76 \text{ V}$

$Zn^{2+}(aq)(\text{cathode}) \rightarrow Zn^{2+}(aq)(\text{anode})$ $\quad\quad\quad\quad\quad\quad \xi° = 0.0 \text{ V}$

$\xi = \xi° - 0.02569/n \times \ln [Zn^{2+}(0.15 \text{ M})]/[Zn^{2+}(1.0 \text{ M})]$

$\xi = 0.0 \text{ V} - ((0.02569/2 \text{ mol e}^-) \times \ln [0.15 \text{ M}/1.0 \text{ M}]) = 0.024 \text{ V}$ **(Nernst equation)**

24. The process of electrolysis occurs in electrolytic cells. The term electrolysis is used to describe any nonspontaneous reaction that only takes place with the addition of electrical energy. **(Electrolysis)**

25. $2Cl^-(l) \rightarrow Cl_2(g) + 2e^-$ (Anode)

    $2Na^+(l) + 2e^- \rightarrow 2Na(l)$ (Cathode)

    $2Na^+(l) + 2Cl^-(l) \rightarrow 2Na(l) + Cl_2(g)$ (Overall reaction) **(Electrolysis)**

26. $2H^+(aq) + 2e^- \rightarrow H_2(g)$ $\xi° = 0.00$ V

    $2H_2O(l) + 2e^- \rightarrow H_2(g) + 2OH^-(aq)$ $\xi° = -0.83$ V **(Electrolysis)**

27. Oxidation occurs at the anode. Thus the two possible oxidations are that of H_2O or $SO_4{}^{2-}$. However, only O_2 is observed at the electrode. This means that $SO_4{}^{2-}$ less readily undergoes oxidation under the conditions of the cell.

    $2H_2O(l) \rightarrow O_2(g) + 4H^+(aq) + 4e^-$ (Anode)

    Reduction occurs at the cathode. The two possible reductions are that of H_2O and Cu^{2+}. Because Cu forms at the cathode, Cu^{2+} undergoes reduction more readily than water. This can easily be determined from their standard reduction potentials.

    $Cu^{2+}(aq) + 2e^- \rightarrow Cu(s)$ $\xi° = +0.32$ V

    $2H_2O(l) + 2e^- \rightarrow H_2(g) + 2OH^-(aq)$ $\xi° = -0.83$ V

    The standard reduction potential for Cu^{2+} is significantly more positive than that for H_2O.

    $2H_2O(l) \rightarrow O_2(g) + 4H^+(aq) + 4e^-$

    $2Cu^{2+}(aq) + 4e^- \rightarrow 2Cu(s)$

    $2Cu^{2+}(aq) + 2H_2O(l) \rightarrow O_2(g) + 4H^+(aq) + 2Cu(s)$ (Overall Reaction) **(Electrolysis)**

28. Anode reaction = $2H_2O(l) \rightarrow O_2(g) + 4H^+(aq) + 4e^-$

    Cathode reaction = $2H_2O(l) \rightarrow H_2(g) + 2OH^-(aq)$

    Overall reaction = $2H_2O(l) \rightarrow 2H_2(g) + O_2(g)$ **(Electrolysis)**

29. a. One ampere is one coulomb per second (1 A = 1 C/s). b. Coulombs (C) = amperes (A) × seconds (s) c. The Faraday constant is the charge on one mole of electrons, 96,500 C/mol e⁻. **(Electrical units)**

30. a. $Pb(s) + Zn^{2+}(aq) \rightarrow Zn(s) + Pb^{2+}(aq)$, b. Reverse the Pb/Pb^{2+} half reaction, change the sign of its standard reduction potential, and add the half reaction to get the desired chemical equations.

    $$Pb(s) + Zn^{2+}(aq) \rightarrow Zn(s) + Pb^{2+}(aq) \quad \xi° = -0.63 \text{ V}$$

    c. This reaction is nonspontaneous because it has a negative potential. Thus, a strip of Pb does not dissolve in a 1 M Zn^{2+} solution. **(Voltaic cells)**

31. $C = A \times s = 10.0$ A × 1.00 min × (60 s/min) = 600 C

    Mol e⁻ = C × (1 mol e⁻/96,500 C) = 600 C × (1 mol e⁻/96,500 C) = 6.22 × 10⁻³ mol e⁻ **(Electrolysis)**

32. $Pb^{2+}(aq) + 2e^- \rightarrow Pb(s)$

    25.0 min × (60 s/min) × (5.00 C/s) × (1 mol e⁻/96,500 C) × (1 mol Pb/2 mol e⁻) × 207 g Pb/mol Pb = 8.04 g Pb **(Electrolysis)**

33. $Fe^{3+}(aq) + 3e^- \rightarrow Fe(s)$

    35.0 g Fe × (1 mol Fe/55.8 g) × (3 mol e⁻/1 mol Fe^{3+}) × (96,500 C/mol e⁻) ×)(1 s/20.0 C) ×
    (Electrolysis) (1 min/60 s) = 151 min

34. a. Fuel cells are voltaic cells in which the substances that undergo oxidation and reduction, the "fuel", come from outside the cell. b. An example of a common type of fuel cell is the one that uses $H_2(g)$ and $O_2(g)$. **(Practical voltaic cells)**

35. a. Mg is the anode because it is a stronger reducing agent than Cd and the anode is the site of oxidation. **(Voltaic cells)**

36. b. Mg^{2+}, Cd^{2+}, the cations, are attracted to the cathode. **(Voltaic cells)**

37. a. Mg is the negative electrode, the anode in a voltaic cell. **(Voltaic cells)**

38. c. $Mg(s)|Mg^{2+}(aq)||Cd^{2+}(aq)|Cd(s)$ The oxidation is written on the left and the reduction is on the right. **(Voltaic cells)**

39. d. F_2 has the highest standard reduction potential because it is the strongest oxidizing agent. **(Reduction potentials)**

40. c. $\xi° = +0.88\ V + 0.49\ V = 1.37\ V$ **(Standard cell potentials)**

Grade Yourself

Circle the numbers of the questions you missed, then fill in the total incorrect for each topic. If you answered more than three questions incorrectly, you need to focus on that topic. (If a topic has less than three questions and you had at least one wrong, we suggest you study that topic also. Read your textbook, a review book, or ask your teacher for help.)

Subject: Electrochemistry

Topic	Question Numbers	Number Incorrect
Electrolytic cells and electrolysis	1, 24, 25, 26, 27, 28, 31, 32, 33	
Votaic cells	2, 5, 30, 35, 36, 37, 38	
Cell notation	3, 4, 15	
Volts and electrical units	6, 29	
Cell potentials	7, 8, 11	
Reduction potentials and standard potentials and standard cell potentials	9, 10, 12, 16, 39, 40	
Cell reaction and potential	13	
Anode and cathode	14	
Cell potential and thermodynamics	17, 18, 19	
Standard potential and equilibrium constant relationship	20	
Nernst equation	21, 22, 23	
Practical voltaic cells	34	

Nuclear and Radiation Chemistry

Brief Yourself

Nuclear chemistry is the study of the nucleus and how it changes. The nucleus is the very small dense region within an atom that contains the nucleons–the protons and neutrons. The nucleus is bonded by the strong or nuclear force, which acts on nucleons separated by less than 10^{-15} m.

The stability of a nuclide depends on its composition. Nuclides with 2, 8, 20, 50, 82, and 126 protons or neutrons are more stable than those with other compositions. Atoms with both an even number of protons and neutrons are generally more stable than those with odd numbers of protons and neutrons. Nuclear stability also correlates with the ratio of neutrons to protons. Small atoms are most stable when they have an equal number of protons and neutrons, while higher-mass atoms are most stable when they have more neutrons than protons. All atoms with an atomic number greater than 83 are radioactive.

Alpha particles are high-energy helium nuclei, $_2^4\text{He}^{2+}$ or $_2^4\alpha$, that have a low penetration power. When a nucleus undergoes α decay, its atomic number decreases by two and its mass number decreases by four. Beta particles are high-energy electrons, $_{-1}^0e^-$ or $_{-1}^0\beta$, that have a penetration power greater than α rays but less than γ rays. During β decay, a neutron changes to a proton releasing a β particle and an antineutrino, $\bar{\nu}$. The daughter nuclide in β decay has the same mass number as the parent and an atomic number one greater than the parent $(Z + 1)$. Gamma rays, γ, are a type of electromagnetic radiation. They have a higher penetration power than either α or β rays. In γ emission, an unstable nucleus releases a γ ray and drops to a lower energy state. Positrons are antielectrons, $_{+1}^0e$ If a positron and electron collide, they annihilate each other and produce two γ rays. In positron emission, a proton changes to a neutron, releasing the positron and a neutrino. The daughter nuclide in positron emission has the same mass number and has an atomic number of $Z - 1$. Electron capture occurs when a proton attracts an electron in a low energy level. This proton changes to a neutron and a neutrino is released. When a higher-energy electron fills the vacated orbital, the atom releases an x-ray.

In nature, elements with mass numbers above 83 undergo radioactive decay in a stepwise fashion to reach more stable nuclear configurations. The three principal natural radioactive decay series begin with ^{238}U, ^{235}U, and ^{232}Th, and end with ^{206}Pb, ^{207}Pb, and ^{208}Pb, respectively.

Scientists can artificially change the nuclear composition of an atom through bombardment reactions. They place a target nucleus into a beam of high-energy particles such as α particles, β particles, neutrons, or protons. Upon bombardment, the target nuclide changes to another nuclide. Under certain conditions, the target nucleus splits and undergoes nuclear fission. Atoms that undergo fission liberate neutrons that can split other nuclei. In addition, a tremendous quantity of energy is released. The opposite of nuclear fission is nuclear fusion. Nuclear fusion results when two low-mass atoms collide and produce a higher-mass atom. Scientists use bombardment reactions to synthesize transuranium elements.

In nuclear chemistry, one of the principal units of energy is the electron volt, eV. One electron volt is the energy needed to accelerate an electron by a potential difference of one volt, and is equivalent to 1.602×10^{-19} J.

The energy needed to separate the nucleons in a nucleus is the nuclear binding energy. Thus, when a nucleus forms it releases the nuclear binding energy. According to Einstein's equation, $\Delta E = \Delta m c^2$, a release of energy, ΔE, has an equivalent loss in mass, Δm. The difference in the sum of the masses of the nucleons and the mass of the nucleus is the mass defect.

Radioactive decay follows first-order kinetics which means that the rate of decay depends on the number of radioactive atoms present. Thus the following equation is used to predict the number of atoms present versus time.

$$\ln\left(\frac{N_0}{N}\right) = kt$$

The half-life, $t_{1/2}$, of radioactive decay is the amount of time for one-half of the nuclei to decay. Half-life relates to the decay constant, k, as follows.

$$t_{1/2} = 0.693/k$$

Photographic film, Geiger counters, scintillation counters, and cloud chambers are often used to detect the presence of radiation.

Many different radiation measurement units are needed. A curie, Ci, is a unit of the activity of a source of radiation. One curie, 1 Ci, is the unit of activity equal to 3.7×10^{10} disintegrations per second, the activity of a 1-g sample of radium. The SI unit of activity is the becquerel, Bq, which is the number of disintegrations per second. Another commonly encountered unit is the roentgen, R, which is a measure of the amount of exposure to x- and γ rays. When one roentgen, 1 R, of either x- or γ rays passes through 1 cm^3 of air at standard conditions, it ionizes air molecules and produces 2.1×10^9 ion pairs. Scientists measure the radiation dosages received by living things and objects. A rad, rd, which stands for radiation absorbed dose, is the amount of radiation that transfers 0.01 J of energy to 1 kg of a specific tissue or substance. The SI unit for radiation absorbed dose is the gray, Gy. One gray, 1 Gy, transfers 1 J of radiation to 1 kg of absorbing substance. Because 1 rad of α radiation can produce more damage than 1 rad of β radiation, scientists multiply a factor called the relative biological effectiveness, *RBE*, times the number of rads to give the effective radiation dosage in rems.

Test Yourself

1. What is the composition of the nucleus of a ^{39}K atom?

2. Compare the N/Z ratios for low-mass nuclei to those of high-mass nuclei.

3. Which of the following has a more stable nucleus? Explain.

    $^{40}_{19}K$ or $^{40}_{20}Ca$

4. Which of the following has a more stable nucleus? Explain.

    $^{209}_{83}Bi$ or $^{209}_{84}Po$

5. Write the general nuclear equation that occurs during alpha decay.

6. Write the equation for the alpha decay of plutonium-242.

7. Write an equation that shows the change that occurs in the nucleus during beta decay. Explain this equation.

8. Write the equation for the beta decay of iodine-131.

9. Describe how gamma decay is different from alpha and beta decay.

10. Explain and write an equation that shows what happens to a nucleus that undergoes electron capture, EC.

11. Write an equation for the positron emission of sodium-22.

12. Compare electron capture and positron emission.

13. Write a nuclear equation for the β decay of $^{227}_{89}Ac$.

14. Write a nuclear equation for the α decay of $^{252}_{100}Fm$.

15. What are the three natural nuclear decay series?

16. When $^{14}_{7}N$ is bombarded with α particles, $^{17}_{8}O$ and a proton are produced. Write a nuclear equation for this reaction.

17. If $^{27}_{13}Al$ is bombarded with a neutron, $^{27}_{12}Mg$ and a proton result. a. Write the nuclear equation for this reaction. b. Write the shorthand notation for the reaction.

18. When $^{238}_{92}U$ is bombarded with neutrons, $^{239}_{93}Np$ and β particles are produced. Write the shorthand notation for this reaction.

19. When $^{238}_{92}U$ is bombarded with deuterium, $^{2}_{1}H$, the unstable daughter nuclide $^{238}_{93}Np$ results, which undergoes β decay to form $^{238}_{94}Pu$. Write equations to show these changes.

20. One way that a uranium-235 nucleus undergoes fission is that it absorbs a slow neutron and produces barium-139 and krypton-94. Write a balanced nuclear equation for this fission reaction.

21. Two tritium atoms, hydrogen-3, undergo a nuclear fusion reaction and produce helium-4 and protons. Write an equation for this nuclear fusion reaction.

22. a. What is the unit electron volt, eV? b. If the charge on an electron is 1.602×10^{-19} C, calculate the number of joules per electron volt.

23. Consider an exothermic reaction such as the combustion of methane:

    $CH_4(g) + 2O_2(g) \rightarrow CO_2(g) + 2H_2O(g)$

What nuclear mass loss results produces −802 kJ?

24. a. What mass loss is found in the following nuclear fusion reaction that occurs in the sun?

$$^3H + {}^2H \rightarrow {}^4He + {}^1n$$

The masses of ^3H and ^2H are 3.01605 u and 2.01410 u, respectively. The masses of ^4He and ^1n are 4.00260 u and 1.00866 u, respectively. b. What energy in MeV is associated with this mass loss? One u is equivalent to 1.6605×10^{-27} kg.

25. What is the nuclear binding energy? Explain.

26. The nucleus of deuterium consists of just a proton (1.00782 u) and a neutron (1.00867 u). The measured mass of the deuterium nucleus is 2.01410 u. a. What is the mass defect of deuterium? b. What is the nuclear binding energy of deuterium in MeV?

27. ^{213}Pb is a β emitter with a half life of 10 min. How many ^{213}Pb atoms remain in a 1.00-mole sample after 2.00 hrs?

28. ^{224}Ra undergoes emission and produces ^{220}Rn as follows:

$$^{224}Ra \rightarrow {}^{220}Rn + \alpha$$

The half life for this nuclear change is 3.64 days. Starting with a 5.00-g sample of ^{224}Ra, what percent of ^{224}Ra remains after 7.00 days?

29. Describe the radiation unit called the curie, Ci.

30. a. What are the units used to measured the amount of radiation absorbed by living things? b. Describe these units and explain how they are related.

31. An archeologist finds a human jaw bone that has 35% the ^{14}C in a modern jaw bone. If ^{14}C has a half life or 5730 yr, calculate the approximate age of the bone.

32. Describe the difference in meaning of the terms *isotopes* and *nuclides*.

33. Two radioactive nuclides of strontium are ^{89}Sr and ^{90}Sr. The half-life for ^{89}Sr is 51 days, and the half-life of ^{90}Sr is 28 years. Which of these two isotopes decays at a faster rate? Explain.

34. Zinc-65 is a positron emitter. Write an equation that shows the decay of zinc-65.

35. What atom has the following characteristics?

$$N = 124, Z = 80$$

a. ^{204}Hg
b. ^{80}Br
c. ^{124}Sb
d. ^{204}Au
e. none of these

36. Which of the following is most stable?

a. $^{200}_{79}Au$
b. $^{200}_{80}Hg$
c. $^{203}_{82}Pb$
d. $^{205}_{83}Bi$
e. all are unstable

37. Complete the following nuclear equation.

$$^{222}_{86}Rn \rightarrow {}^{218}_{84}Po + \underline{}$$

a. $^4_2\alpha$
b. $^0_{-1}\beta^-$
c. γ
d. ^4H
e. none of these

38. Complete the following nuclear equation:

$$^{238}U + ^{22}Ne \rightarrow \underline{} + 4^1n$$

a. ^{256}Md
b. ^{259}No
c. ^{259}No
d. ^{256}No
e. none of these

39. Which of the following is the shorthand notation for the following nuclear change?

$$^{238}U + ^{16}O \rightarrow ^{249}Fm + 5^1n$$

a. $^{238}U(^{16}O, ^{249}Fm)5^1n$
b. $^{16}O(^{238}U, ^{249}Fm)5^1n$
c. $^{238}U(^{16}O, 5^1n)^{249}Fm$
d. $^{16}O(^{238}U, 5^1n)^{249}Fm$
e. none of these

40. What nuclide results after ^{214}Pb undergoes two successive β emissions?

a. ^{214}Pb
b. ^{214}Hg
c. ^{210}Hg
d. ^{214}Po
e. none of these

Check Yourself

1. The mass number is 39, which means it has a total of 39 protons and neutrons. Its atomic number is 19; hence, it has 19 protons. Subtracting 19 from 39 gives 20 neutrons. **(Nuclear composition)**

2. In nuclides with low mass numbers ($A < 41$), the most stable nuclides are usually those with the same number of protons and neutrons ($N/Z = 1$); e.g., $^{12}_{6}C$, $^{16}_{8}O$, $^{32}_{16}S$, and $^{40}_{20}Ca$. For higher-mass nuclides ($A < 41$) this is not true. The most stable nuclides have a larger number of neutrons than protons ($N/Z > 1$). **(Nuclear stability)**

3. ^{40}Ca has an even number of protons and neutrons, while ^{40}K has an odd number of nucleons. Nuclides that have an even number of nucleons are more stable than those that do not. **(Nuclear stability)**

4. ^{209}Bi has 83 protons and 126 neutrons. Whereas, ^{209}Po has 84 protons and 125 neutrons. Each of these nuclides has one even and one odd N or Z value. However, this is unimportant because the atomic number of ^{209}Po exceeds 83 and all such nucleons are radioactive. **(Nuclear stability)**

5. The general nuclear equation that shows what happens during alpha decay is

$$^{A}_{Z}X \rightarrow ^{A-4}_{Z-2}Y + ^{4}_{2}He$$

in which $^{A}_{Z}X$ is the parent nuclide, $^{A-4}_{Z-2}Y$ is the daughter nuclide, and $^{4}_{2}He$ is an alpha particle. **(Alpha decay)**

6. $^{242}_{94}Pu \rightarrow ^{238}_{92}U + ^{4}_{2}\alpha$ **(Alpha decay)**

7. $^{1}_{0}n^0 \rightarrow ^{1}_{1}p^+ + ^{0}_{-1}e + \bar{\nu}$ Beta decay occurs in nuclei that have a high neutron to proton ratio. Neutrons can undergo a spontaneous change, producing a beta particle, $^{0}_{-1}e^1$; a proton, $^{1}_{1}p^+$; and an antineutrino, $\bar{\nu}$. **(Beta decay)**

8. $^{131}_{53}I \rightarrow ^{131}_{54}Xe + ^{0}_{-1}e + \bar{\nu}$ **(Beta decay)**

9. Gamma emission takes place without a measurable nuclear mass change because γ radiation is a type of electromagnetic energy. After γ emission, the nucleus reaches a lower energy state. In other words, the

same nucleus is present but is more stable. In both alpha and beta decay, a new nucleus results. (**Gamma decay**)

10. During electron capture, a proton in the nucleus captures an electron, usually one in the lowest energy level. This proton changes to a neutron and a neutrino is released. (**Electron capture**)

$$_1^1p^+ + _{-1}^0e^- \rightarrow + _0^1n + \nu$$

11. $_{11}^{22}Na \rightarrow _{10}^{22}Ne + _{+1}^0e^+ + \nu$ (**Positron emission**)

12. Both electron capture and positron emission produce the same nuclear transformation (a proton changes to a neutron), and in many cases they compete with each other. Electron capture occurs more often in high-mass nuclides because the lowest energy orbitals have small radii, which makes it easy for the nucleus to capture an electron. (**Positron emission and electron capture**)

13. $_{89}^{227}Ac \rightarrow _{90}^{227}Th + _{-1}^0e^-$ (**Beta decay**)

14. $_{100}^{252}Fm \rightarrow _{98}^{248}Cf + _2^4He$ (**Alpha decay**)

15. 1. ^{238}U series: ^{238}U to ^{206}Pb

    2. ^{235}U series: ^{235}U to ^{207}Pb

    3. ^{232}Th series: ^{232}Th to ^{208}Pb (**Natural decay series**)

16. $_7^{14}N + _2^4\alpha \rightarrow _8^{17}O + _1^1p^+$ (**Bombardment reactions**)

17. a. $_{13}^{27}Al + _0^1n \rightarrow _{12}^{27}Mg + _1^1p^+$

    b. $^{27}Al(n, p)^{27}Mg$ (**Bombardment reactions**)

18. $_{92}^{238}U + _1^2H \rightarrow _{93}^{238}Np + 2\,_0^1n$ $^{238}U(^1n^0, _{-1}e)^{239}Np$ (**Bombardment reactions**)

19. $^{238}U(^2H, 2\,^1n^0)^{238}Np$

    $^{238}Np \rightarrow ^{238}Pu + °_{-1}e^-$ (**Bombardment reactions**)

20. $_{92}^{235}U + _0^1n^0 \rightarrow _{56}^{139}Ba + _{36}^{94}Kr + 3\,_0^1n^0$ (**Nuclear fission**)

21. $^3H + ^3H \rightarrow ^4He + 2\,^1H$ (**Nuclear fusion**)

22. a. One electron volt is the energy needed to accelerate an electron by a potential difference of one volt.

    b. $1\text{ eV} = C \times 1V = 1.602 \times 10^{-19}\text{ C} \times 1\text{ J/C} = 1.602 \times 10^{-19}\text{ J}$

    The product of the charge in coulombs times the voltage is the number of joules. (**Nuclear units**)

23. $\Delta E = \Delta m\,c^2$

    $-802\text{ kJ} \times (1000\text{ J}/1\text{ kJ}) = m \times (3.00 \times 10^8\text{ m/s})^2$

    $\Delta m = -8.91 \times 10^{-12}\text{ kg} = -8.91 \times 10^{-9}\text{ g} = -8.91\text{ ng}$ (**Energy-mass relationship**)

24. a. $\Delta m = \text{mass}(^4He + ^1n) - \text{mass}(^3H + ^2H)$

    $\Delta m = (4.00260\text{ u} + 1.00866\text{ u}) - (3.01605\text{ u} + 2.01410\text{ u}) = -0.01889\text{ u}$

    b. $\Delta m = -0.01889\text{ u} \times (1.6605 \times 10^{-27}\text{ kg}/1\text{ u}) = -3.137 \times 10^{-29}\text{ kg}$

    $\Delta E = \Delta m\,c^2 = -3.137 \times 10^{-29}\text{ kg} \times (2.998 \times 10^8\text{ m/s})^2 = -2.819 \times 10^{-12}\text{ J}$

    $\Delta E = -2.819 \times 10^{-12}\text{ J} \times 1\text{ eV}/1.602 \times 10^{-19}\text{ J} = -1.760 \times 10^7\text{ eV}$

    $\Delta E = -1.760 \times 10^7\text{ eV} \times (1\text{ MeV}/1 \times 10^6\text{ eV}) = -17.60\text{ MeV}$ (**Energy-mass relationship**)

25. When nucleons combine to form a nucleus, they release energy just as atoms release energy when they form chemical bonds. In other words, the nucleus is more stable than the individual particles from which it is made. The energy needed to separate the nucleons in a nucleus is the nuclear binding energy. **(Nuclear binding energy)**

26. a. Total mass = 1.00782 u + 1.00867 u = 2.01649 u

   Δm = actual mass of nucleus − sum of the masses of the nucleons

   Δm = 2.01410 u − 2.01649 u = −0.00239 u

   b. Δm = −0.00239 u × (1.6605 × 10^{-27} kg/1 u) = −3.97 × 10^{-30} kg

   $\Delta E = \Delta m\, c^2$ = −3.97 × 10^{-30} kg × (2.998 × 10^8 m/s)^2 = −3.57 × 10^{-13} J

   ΔE = −3.57 × 10^{-13} J × (1 eV/1.602 × 10^{-19} J) = −2.23 × 10^6 eV = −2.23 MeV **(Nuclear binding energy)**

27. $t_{1/2} = 0.693/k$

   $k = 0.693/t_{1/2}$ = 0.693/10 min = 0.0693 min^{-1}

   $\ln(N_0/N) = k\,t$

   $\ln(1.00\text{ mol}/N)$ = 0.0693 min^{-1} × 120 min

   N = 2.44 × 10^{-4} mol × (6.022 × 10^{23} atoms/mol) = 1.47 × 10^{20} atoms   **(Kinetics of decay)**

28. $k = 0.693/t_{1/2}$ = 0.693/3.64 d = 0.190 d^{-1}

   $\ln(\text{g }^{224}\text{Ra initial}/\text{g }^{224}\text{Ra final}) = kt$

   $\ln(5.00\text{ g}/\text{g }^{224}\text{Ra final})$ = 0.190 d^{-1} × 7.00 d

   g ^{224}Ra final = 1.32 g

   % ^{224}Ra remaining = (g ^{224}Ra remaining/g ^{224}Ra initial) × 100

   % ^{224}Ra remaining = (1.32 g/5.00 g) × 100 = 26.4%   **(Kinetics of decay)**

29. A curie, Ci, is a unit of the activity of a radiation source. One curie, 1 Ci, is the unit of activity that is equivalent to 3.7 × 10^{10} disintegrations per second, the activity of a 1-g sample of ^{226}Ra. **(Radiation units)**

30. a. The rad, gray, and rem are the three units used most often to measure radiation dosages.

   b. A rad, rd, which stands for radiation absorbed dose, is the amount of radiation that transfers 0.01 J of energy to 1 kg of a specific tissue or substance. The SI unit for radiation absorbed dose is the gray, Gy. One gray, 1 Gy, is the quantity of α radiation that transfers 1 J to 1 kg of absorbing substance. Thus, 1 Gy is equivalent to 100 rd. Because 1 rad of radiation can produce more damage than 1 rad of β radiation, scientists multiply a factor called the relative biological effectiveness, RBE, times the number of rads to give the effective radiation dosage in rems.   **(Radiation units)**

31. $k = 0.693/t_{1/2}$ = 0.693/5730 yr = 1.21 × 10^{-4} yr^{-1}

   $N = 0.35\, N_0$

   $\ln(N_0/N) = kt$

   $\ln(N_0/0.35\, N_0)$ = *1.21 × 10^{-4} yr^{-1}* × t

   t = 8.7 × 10^3 yr   **(Radiocarbon dating)**

32. Isotopes are atom with the same atomic number but have different mass numbers. Nuclides are nuclei that differ in both atomic and mass numbers. They are nuclei of different elements.   **(Isotopes)**

33. ^{90}Sr is decaying at a slower rate because it has a longer half life (28 yr). ^{89}Sr only requires 51 days to decay to one-half of the original sample.   **(Half life)**

34. $^{65}_{30}\text{Zn} \rightarrow\ ^{65}_{29}\text{Cu} +\ ^{0}_{1}e^{+} + v$ **(Positron emission)**

35. a. ^{204}Hg has an atomic number, Z, of 80 and mass number, A, of 204 (80 + 124). (**Nuclide**)

36. b. $^{200}_{80}$Hg is the only one that has an even number of nucleons; thus, it has the most stable nucleus. (**Stability of nuclei**)

37. a. $^{4}_{2}$He an alpha particle balances the mass and atomic numbers. (**Nuclear equations**)

38. d. ^{256}No (**Nuclear equations**)

39. c. ^{238}U(^{16}O, 5^{1}n)^{249}Fm (**Nuclear equations**)

40. d. ^{214}Pb $\rightarrow$ ^{214}Bi + $_{-1}\beta$
    ^{214}Bi $\rightarrow$ ^{214}Po + β (**Beta decay**)

Grade Yourself

Circle the numbers of the questions you missed, then fill in the total incorrect for each topic. If you answered more than three questions incorrectly, you need to focus on that topic. (If a topic has less than three questions and you had at least one wrong, we suggest you study that topic also. Read your textbook, a review book, or ask your teacher for help.)

Subject: Nuclear and Radiation Chemistry

Topic	Question Numbers	Number Incorrect
Nuclear composition	1	
Nuclear stability, kinetics of decay, half life, and stability of nuclei	2, 3, 4, 27, 28, 33, 36	
Alpha decay	5, 6, 14	
Beta decay	7, 8, 13, 40	
Gamma decay	9	
Electron capture	10, 12	
Positron emission	11, 12, 34	
Natural decay series	15	
Bombardment reactions	16, 17, 18, 19	
Nuclear fission	20	
Nuclear fusion	21	
Nuclear units	22	
Energy-mass relationship	23, 24	
Nuclear binding energy	25, 26	
Radiation units	29, 30	
Radiocarbon dating	31	
Isotopes	32	
Nuclide	35	
Nuclear equations	37, 38, 39	

RECEIVED
SEP 21 2001